The Law of the Whale Hunt

Whale oil lit the cities and greased the machines of the Industrial Revolution. In light of its importance, competition between whalers was high. Far from courts and law enforcement, competing crews of American whalers not known for their gentility and armed with harpoons tended to resolve disputes at sea over ownership of whales. Left to settle arguments on their own, whalemen created norms and customs to decide ownership of whales pursued by multiple crews.

The Law of the Whale Hunt provides an innovative examination of how property law was created in the absence of formal legal institutions regulating the American whaling industry in the eighteenth and nineteenth centuries. Using depositions, court testimony, logbooks, and other previously unused primary sources, Robert Deal tells an exciting story of American whalers hunting in waters from the North Atlantic to the South Pacific and the Sea of Okhotsk.

ROBERT DEAL is an assistant professor of history at Marshall University, West Virginia.

Cambridge Historical Studies in American Law and Society

Series editor

Christopher Tomlins, *University of California, Berkeley*

Previously published in the series:

Sandra F. VanBurkleo, Gender Remade: Citizenship, Suffrage, and Public Power in the New Northwest, 1879–1912
Reuel Schiller, Forging Rivals: Race, Class, Law, and the Collapse of Postwar Liberalism
Ely Aaronson, From Slave Abuse to Hate Crime: The Criminalization of Racial Violence in American History
Stuart Chinn, Recalibrating Reform: The Limits of Political Change
Ajay K. Mehrotra, Making the Modern American Fiscal State
Yvonne Pitts, Family, Law, and Inheritance in America: A Social and Legal History of Nineteenth-Century Kentucky
David M. Rabban, Law's History
Kunal M. Parker, Common Law, History, and Democracy in America, 1790–1900
Steven Wilf, Law's Imagined Republic
James D. Schmidt, Industrial Violence and the Legal Origins of Child Labor
Rebecca M. McLennan, The Crisis of Imprisonment: Protest, Politics, and the Making of the American Penal State, 1776–1941
Tony A. Freyer, Antitrust and Global Capitalism, 1930–2004
Davison Douglas, Jim Crow Moves North
Andrew Wender Cohen, The Racketeer's Progress
Michael Willrich, City of Courts, Socializing Justice in Progressive Era Chicago
Barbara Young Welke, Recasting American Liberty: Gender, Law and the Railroad Revolution, 1865–1920
Michael Vorenberg, Final Freedom: The Civil War, the Abolition of Slavery, and the Thirteenth Amendment
Robert J. Steinfeld, Coercion, Contract, and Free Labor in Nineteenth Century America
David M. Rabban, Free Speech in Its Forgotten Years
Jenny Wahl, The Bondsman's Burden: An Economic Analysis of the Common Law of Southern Slavery
Michael Grossberg, A Judgment for Solomon: The d'Hauteville Case and Legal Experience in the Antebellum South

The Law of the Whale Hunt

Dispute Resolution, Property Law, and American Whalers, 1780–1880

ROBERT DEAL
Marshall University

CAMBRIDGE
UNIVERSITY PRESS

CAMBRIDGE
UNIVERSITY PRESS

32 Avenue of the Americas, New York, NY 10013–2473, USA

Cambridge University Press is part of the University of Cambridge.

It furthers the University's mission by disseminating knowledge in the pursuit of education, learning, and research at the highest international levels of excellence.

www.cambridge.org
Information on this title: www.cambridge.org/9781107114630

First published 2016

Printed in the United States of America

A catalog record for this publication is available from the British Library.

Library of Congress Cataloging in Publication Data
Deal, Robert, 1958– author.
The law of the whale hunt : dispute resolution, property law, and American whalers, 1780–1880 / Robert Deal, Marshall University.
pages cm
ISBN 978-1-107-11463-0 (Hardback)
1. Whaling–Law and legislation–United States–History–19th century. 2. Whaling–United States–History–19th century. 3. Dispute resolution (Law)–United States–History–19th century. 4. Property–United States–History–19th century. I. Title.
KF1770.D43 2015
343.7307′692809034–dc23 2015031738

ISBN 978-1-107-11463-0 Hardback

For Liz

Contents

Figures *page* ix
Acknowledgments xi

1 Introduction 1
2 The biology, economics, and geography of Anglo-American whaling 14
3 Fast-fish, loose-fish 29
4 Iron holds the whale 58
5 Laws of honor 82
6 Halves 106
7 The environment of litigation 133
8 An untragic commons 144
9 Conclusion 159

Index 173

Figures

1 Greenland whale; Sperm whale. *Naturgeschichte der Säugethiere: mit colorirten Abbildungen zum Anschauugs-Unterricht für die Jugend*, 1872. *page* 15
2 Bowhead whale head on deck. 19
3 Baleen in warehouse of William Lewis and Son. 19
4 Toggle iron. 24
5 "Trying Out." Harper's Magazine, 1874. 26
6 Tryworks at full blast. 27
7 Map of the North Polar Regions, 1882. 30
8 Rev.d William Scoresby, 1824. 31
9 Vessel in ice with icebergs behind at left. 32
10 "Grand Ball Given by the Whales in Honor of Oil Wells in Pennsylvania." Vanity Fair, 1861. 157

Acknowledgments

This book began as a doctoral dissertation in the history department at Temple University. I must confess that I entered graduate school without ever having given whales or whaling any thought. My advisor, Andrew Isenberg, suggested whales as an area of study and lacking any other ideas I took on the topic. Drew probably never expected a project that started out as an environmental history of whaling to quickly become an exercise in legal history. I am extremely grateful that he allowed and encouraged me to take the dissertation off in the direction that made the most sense to me. Drew was a superb advisor and mentor and my debt to him is large. The other members of my dissertation committee – David Waldstreicher, Elizabeth Varon, and Harwell Wells – were extremely helpful and encouraging. It is difficult to imagine a finer group of scholars and people to work with. The history department at Temple University was a remarkably pleasant and supportive place to study, and I have nothing but thanks and praise for the faculty and my fellow graduate students.

As this project has progressed I have accumulated a great many debts. I would like, in particular, to thank the staffs of the following institutions for their assistance: The New Bedford Whaling Museum Library (New Bedford, MA), The John Carter Brown Library (Providence, RI), Baker Library Historical Collections (Cambridge, MA), Houghton Library (Cambridge, MA), National Archives (Waltham, MA), Hawaii State Archives (Honolulu), New York Public Library, and the Hull Maritime Museum (Kingston upon Hull, England). Portions of this work have been presented at a number of different conferences and venues. I have benefited greatly from the comments, criticisms, and suggestions offered at the Society of Historians of the Early American Republic Conference, the Ohio Valley History Conference, University of Toronto Faculty of Law, Law and Humanities Junior Scholar Workshop, World History Association Conference, the Harvard Graduate Student Conference on International History, and at Final Frontiers: Exploring Oceans,

Islands, and Coastal Environments sponsored by the Rachel Carson Center for Environment and Society (Munich) and the Island Institute (Rockland, ME).

I have also benefited greatly from the encouragement, comments, assistance, and suggestions of many individuals. I would particularly like to thank: Bryant Simon, Drew Isenberg, Elizabeth Varon, David Waldstreicher, Harwell Wells, Angela Fernandez, Bruce Ziff, Robert C. Ellickson, Christopher Tomlins, David Armitage, Arthur F. McEvoy, Carol Rose, Robert Gordon, John Gillis, Franziska Torma, Jeffrey Bolster, Michael Dyer, Mark Procknik, Travis Roy, Joshua Parsons, Chris Calton, several anonymous reviewers, the many students I have taught at Marshall University, and – long before this project was ever conceived – Meredith Parsons Lillich who instilled in me the idea that I might just have the aptitude for this sort of work.

I would like to thank all of the people at Cambridge University Press and particularly series editor Christopher Tomlins for championing this book. It is a much better book because of his involvement.

I must admit that I always viewed the language stating that an author thanks contributors but is responsible for any errors as boilerplate. It is only now – having gone through this process – that I understand just how heartfelt that sentiment is.

Portions of this book first appeared as: "The Judicial Invention of Property Norms: Ellickson's Whalemen Revisited," *University of Toronto Law Journal*, Vol. 63, Issue 1 (2013), 73–96. (DOI:10.3138/utlj.63.1deal). Reprinted with permission from University of Toronto Press (www.utpjournals.com), © University of Toronto Press 2013. All rights reserved; and as "Fast Fish, Loose-Fish: How Whalemen, Lawyers, and Judges Created the British Property Law of Whaling" originally published in the *Ecology Law Quarterly*, 37:2 (2010). I am very grateful to the University of Toronto Press and the Regents of the University of California for granting permission to use portions of those articles in this book.

I have also accumulated a great many debts of a more personal nature, and I would very much like to acknowledge the support offered by colleagues, friends, and family. My colleagues in the history department at Marshall University have been unfailingly supportive of my work, and I am forever grateful that they took a chance and hired a rather nontraditional candidate. Dan Holbrook, as the Chair of the History Department since my arrival at Marshall in 2010, has eased my transition into academia and teaching. I also appreciate the support of the larger community of scholars at Marshall University. Many friends and family have provided a welcome diversion from work and so have also added much to this project. I would particularly like to acknowledge my brothers Bruce Deal and Bill Deal, Waite Stuhl, Joan Halbert, Beth Zimmerman, Phil Rutherford, Laura Michele Diener, Kat Williams, Greta Rensenbrink, Michael Woods, Margie Hutzayluk, Dave Hutzayluk, Mary Virginia Heyse, Jim Hoffman, and Liz Hoffman.

The one sorrow of a first book published deep into one's middle years is that friends and family members one would like to thank are no longer around. In particular, my parents, Gerry and Clarence Deal, were always supportive and I am confident that they would have been proud of this book. The joy, however, of a first book published on the cusp of sixty is that I can thank new family members – Kate Deal, Jason Kelly, and Aiden Kelly – for their support and the pleasure of their company.

I would like to thank my children – Christopher, Beverly, and Lydia – for their love, support, and understanding of why going to graduate school and pursuing an academic career was so important to me. I only hope that they are half as proud of me as I am of them.

My wife Liz has been my partner and the great love of my life for nearly forty years, and I owe more to her than I even understand or know how to express. This book is for her.

1

Introduction

This book examines how American whalemen resolved disputes at sea over the possession of whales. It also looks at the development of the law governing such disputes to understand how property law is created. Legal scholars and historians have tended to frame the issue of property law creation as an either/or question. Is it produced by legislators and judges, or does it develop from the practices and customs of involved individuals? To insist that property law must be the product of one or the other is to impose consistency and order on a process that is messy and often haphazard. Property law – like most things people produce collectively – emerges from a mélange of misunderstandings, mistakes, and contradictions. Law reflects both greed and a sincere desire to find fair solutions to difficult problems. It is never entirely one thing or the other. What follows is an examination of how Anglo-American whalemen, lawyers, judges, and legal scholars created the laws governing property disputes over contested whales in the period from 1780 to 1880.

There was never a single property law of whaling or even a single set of rules operating at sea to settle arguments over the possession of a successfully hunted whale. Whalemen operated at sea according to a number of general maxims that were often poorly understood even by experienced captains and crews. Negotiations and a sense of what constituted proper behavior – what one British captain termed "laws of honor" – were ultimately more important in resolving disputes over contested whales than any universally honored custom. What whalemen valued most was a means of dispute resolution that within a framework of honorable behavior prevented violence and facilitated the killing of whales. When conflicts came before Anglo-American courts, lawyers and judges had different concerns. Recognizing that whaling vessels flew the flags of competing nations, jurists were worried that a simple argument between rival captains might develop into an international incident. Adopting industry customs respected by all whalemen seemed to offer a neutral means of resolving cases that did not honor the law of one nation over another. In their effort to discern whaling customs, courts often misunderstood how whalemen actually settled disputes and found universal practices when, in fact, a flexible and ad hoc means of awarding whales existed at sea. The law of whaling that emerged from the reported Anglo-American cases must – in its clear statement of applicable customs and evocation of arguments dating back to Justinian about how one comes to possess and own wild animals – have seemed quite foreign to the men who made their living hunting whales. Whalemen, in turn, largely ignored judicial pronouncements as to the customs of whaling and continued to operate in ways that made sense to them in their relentless quest to kill whales.[1]

The primary reason Anglo-American courts failed to understand how whalemen settled disputes is that lawyers and judges were really never all that interested in or concerned about whaling practices. Although whaling was an important industry in the eighteenth and nineteenth centuries, it was always something of a legal backwater. The participants formed a discrete community engaging in an isolated and unique activity. Judges never worried that a ruling in a whaling case might upset established laws followed by other maritime industries. Courts in seeking settled whaling customs and universal adherence largely invented what they were looking for.

In addition, when the legal profession did think about whaling, members of the bench and bar failed to grasp that whales are extremely large and difficult to catch. Even in the late eighteenth and early nineteenth centuries when sperm whales and bowheads were plentiful, the finding and catching of a whale was time consuming and extremely dangerous. Weeks or months might pass in some seasons before a ship reduced a whale to oil. The catching of a single additional whale might be the difference between a successful season and financial loss. The whaling industry operated in almost all waters and time periods on a model of scarcity. Unlike bison hunting, logging, salmon fishing, or other extractive industries of the nineteenth century, whalemen never had the luxury of passing up a whale with the confidence that other cetaceans could be easily located and killed. The happy discovery of a large number of whales swimming in company was no guarantee that even a single animal would be taken. Courts, when thinking about wild animal cases, imagined ducks, rabbits, mackerel, or even bees. The stakes in a dispute over a single animal worth potentially thousands of dollars bore little resemblance to a contest concerning a small animal that individually was of little value.

When the writers of Anglo-American legal treatises in the nineteenth century took up the task of explaining property rights in wild animals – or what the law calls *ferae naturae* – they never questioned whether the reported cases provided an accurate description of whaling practices. Their concern was in drawing

[1] C. Ian Jackson, ed., *The Arctic Whaling Journals of William Scoresby the Younger: Volume I, The Voyages of 1811, 1812 and 1813* (London: The Hakluyt Society, 2003), 119–20.

larger principles from these matters that would be applicable in other property cases. Whales, deer, and foxes were all – in the eyes of the law – pretty much the same. Even if these scholars suspected that whaling customs were not universally observed, it would have likely made little difference in their explication of the law. In common law jurisdictions, lawyers accept the statement of facts set forth in an opinion as proven. The ruling which follows is predicated on those facts. Facts in an opinion are thereafter stretched and manipulated by lawyers and judges to discover the point at which different facts necessitate a different ruling.

While the legal profession of the nineteenth century did much to obscure actual whaling customs, it has been the misreading of Herman Melville's discussion of whaling law in *Moby-Dick* that has enshrined the idea in the minds of most historians that whalemen avoided litigation and violence by a strict adherence to universally accepted norms. Melville proclaimed that without such universal laws, "vexatious and violent disputes" would frequently arise between ships claiming ownership of a whale. Melville reduced this unwritten property law of whaling to a pair of pithy maxims. "I. A Fast-Fish belongs to the party fast to it. II. A Loose-Fish is fair game for anybody who can soonest catch it." Yet, as one reads on and Melville's explanation of whaling laws becomes more convoluted, it becomes clear that these seemingly simple principles were – in practice – neither well understood nor universally followed at sea.[2]

Like most attempts at brevity and concision in the law, Melville's summation raised more questions than it answered. What, for example, constituted a fast-fish? How much control must a whaler have had over his prey before it was deemed fast? Melville's gloss provided some answers. A whale was fast when it was connected to an occupied boat by anything within the control of the crew. "[A] mast, an oar, a nine-inch cable, a telegraph wire, or a strand of cobweb, it is," Melville declared, "all the same." In Melville's telling, control of a whale's fate or even its movement was clearly not required. The fictive nature of control in obtaining the right to a whale was emphasized by Melville's further explanation that a whale was also "technically fast" when it carried the waif or "other recognized symbol of possession" of a ship that had the present ability and intention of taking the animal.[3]

Although Melville presented whaling norms as universal and unchanging, his description of these practices in *Moby-Dick* captured much of the confusion and ambiguity that governed confrontations at sea. Melville managed to seamlessly conflate two standards that would, if strictly applied, render contradictory results. Fast-fish, loose-fish, which Melville deemed the universal law, required that a physical connection between whale and boat or crew member must be maintained to defeat the claims of a rival vessel. Yet, he also introduced the norm of iron holds the whale which provided that a boat retained its claim

[2] Herman Melville, *Moby-Dick or, The Whale* (New York: Penguin Books, 2003), 432–3.

[3] Melville, *Moby-Dick*, 433. A waif is a flag that is attached to a pole that is affixed to a dead whale.

to a whale even in the absence of an attached line if a properly marked harpoon remained fast and the ship continued in pursuit with the ability – absent interference – to capture its prey. Melville indicated that a third standard – justice – was also sometimes invoked by the more honorable whalemen to award whales to captains whose claims, while ethically compelling, were weak under the prevailing norms.[4]

How then did whalemen use such vague guidelines to settle arguments at sea? A review of the available evidence concerning disputes at sea between whalemen vying for the same quarry reveals that whaling customs were often vague and subject to interpretation and negotiation. While Anglo-American whalemen – if asked to state the applicable custom – would have agreed that the first boat to affix a harpoon gained an advantage over its competitors, the precise application of this principle was far from clear. Whaling customs during the eighteenth and nineteenth centuries were, in fact, a jumble of often competing maxims tied together by what Melville saw as the desire of "the more upright and honorable whalemen" to do justice. Yet, without resorting to violence, whalemen managed to resolve disputes at sea because they were a close-knit community that shared both a commitment to harvesting the maximum amount of oil and bone and ideas as to what constituted fairness.[5]

The ability of whalemen to resolve disputes on their own without violence and without recourse to lawyers and judges was a remarkable achievement made possible by the community's tight social structure. Anglo-American whalemen were almost always members of close-knit communities. The British whaling industry in the Greenland fishery of the late eighteenth and early nineteenth centuries was primarily based in a small number of ports on England's northeast coast and in nearby Scottish towns. Captains and crews were often well acquainted with the men on other vessels. Even when British and American ships began competing during the last decades of the eighteenth century for whales in the southern hemisphere, a high percentage of captains and crews – regardless of the ship's flag – hailed from Nantucket. In the American dominated nineteenth-century fisheries of the Pacific Ocean and, after 1848, the Western Arctic, virtually all of the ships called New Bedford, Sag Harbor, or one of a handful of Southern New England towns home. Identifying all of the bonds of marriage and consanguinity between American ship owners, captains, officers, and crews would be a daunting task. The tightness of this community was increased by the tendency of American whalers by the 1820s to spend significant periods each winter in Honolulu and other Hawaiian ports. There the men worshipped, socialized, and replicated – to a limited degree – New England society.[6]

[4] Melville, *Moby-Dick*, 432–5. [5] Melville, *Moby-Dick*, 433–5, 443.

[6] Gordon Jackson, *The British Whaling Trade* (Hamden, CT: Archon Books, 1978); Eric Jay Dolin, *Leviathan: The History of Whaling in America* (New York: W. W. Norton & Co., 2007), 212–5; and Robert C. Ellickson, *Order without Law: How Neighbors Settle Disputes* (Cambridge: Harvard University Press, 1991), 192–5.

Legal theorists and economic historians have generally accepted Melville's notion of universally observed whaling customs and elided his confusing and contradictory discussion of those practices. Like the writers of legal treatises in the nineteenth century, recent scholars have used the handful of reported whaling cases as a vehicle for making larger points about the development of property law. Actual whaling practices have received very limited scrutiny and whalemen have been largely cast as wealth or welfare maximizing automatons at the service of the so-called norms scholars who see property law as largely the bottom-up creation of communities of users.[7]

To understand the internecine struggles between the norms school and the legal centralists who view property law as the work of legislatures and other legal elites, the work of the economist Ronald Coase is central. Coase, a sort of latter day Hobbesian and a leading legal centralist, believed – as did most of his contemporaries – that property law was imposed from above by courts and legislators. While Coase did not, as we shall see, believe that it ultimately much mattered what the law dictated, he certainly never envisioned participants creating the governing rules. In 1960, Coase published an article setting forth an extremely influential model of how people with competing interests in property and natural resources settle disputes. Coase illustrated what has come to be known as the Coase Theorem with a hypothetical dispute involving a rancher whose cattle destroys the crops of the neighboring farmer.[8]

If the law in this fictional jurisdiction holds the rancher liable for damage inflicted by his cattle to the farmer's crops, Coase envisioned several outcomes. The rancher could install a fence and add as many animals to the herd as is practicable. However, the rancher, realizing that the cost of the fence is greater than the harm inflicted on the farmer, may decide that it makes more economic sense to pay the farmer for the damage to the crops. This would be acceptable to the rancher so long as the income generated by a particular animal exceeds the value of the damaged crop. Likewise, the farmer would be satisfied if the money received from the rancher is greater than the loss inflicted by the wayward ungulates. In a situation where the rancher and farmer are both satisfied, the rancher will feel free to add animals to his herd, confident that his profits will rise. The end result will be more cattle, a diminished crop yield, and an improvement in the bottom line for both neighbors. If, however, the fence does not make economic sense and the value of the crop damage caused by a single animal exceeds the profit generated by that creature, the rancher will not add to his herd.[9]

[7] Ellickson, *Order without Law*, 191–206; and Henry E. Smith, "Community and Custom in Property," *Theoretical Inquiries in Law* 10, no. 1 (January 2009): 6.

[8] Ronald Coase, "The Problem of Social Cost," *Journal of Law & Economics* 3 (October 1960), 1–44.

[9] Coase, "The Problem of Social Cost," 1–8. The crop yield will be diminished because of the damage or by way of an agreement between the neighbors that the farmer will leave the field vacant. Another possibility is that the farmer will choose to grow a crop that cattle do not harm.

Coase then asked what would happen if the law were changed and the rancher was no longer responsible for damage caused by his cattle. The farmer will build a fence if the cost does not exceed the value of the damaged crops. The same result is thus reached as in the system where the rancher is liable. So long as crop damage is greater than the cost of the fence, the fence will get built and the rancher will add to his herd as he sees fit. The only difference between the two legal systems is the party who absorbs the cost of the fence. Assuming that the farmer does not build a fence, he will pay the rancher to reduce his herd by an amount that is less than the amount of damage done by that particular animal. The rancher will be happy not to raise the animal if the payment from the farmer is greater than the value of the cow. If, however, the value of one animal is greater than the value of the crops it damages, the rancher will not accept the payment and add to his herd as he chooses. Again, as in the scenario where the rancher is liable, when the value of the cattle is greater than the crop damage the herd will increase.[10]

The Coase Theorem thus demonstrates that the end result of crop and meat to market is identical regardless of upon whom liability for wayward cattle is imposed and upon whom the cost is placed. The importance of Coase's findings was immediate to both lawyers and economists. Liability laws had nothing to do with the amount and type of product reaching the market. The parties would reach the same results regardless of whom the law declared responsible. This was an obvious blow to the notion that liability laws had a major impact upon society at large. The only people with any interest in the matter were the rancher and the farmer. Urban consumers were not affected one way or the other. While Coase overturned the prevailing notion that the content of a law imposing liability necessarily affected production, he hewed to the orthodox view that law is imposed from above.[11]

The Coase Theorem is based upon two very large assumptions. The first is that people always behave in a rational way that is based solely upon economic self-interest. The second assumption is that there are no transaction costs in the negotiations between the rancher and the farmer. Transaction costs are the expenses that parties incur reaching a resolution to their dispute. Typical transaction costs for any dispute arise in preparing for and conducting negotiations and, when resolution proves difficult, litigation. Coase was well aware that human behavior is not simply the product of an accountant's calculations. He also recognized that transaction costs can be significant in shaping how property disputes are settled. Instead, Coase sought to challenge prevailing ideas and suggest that assignment of liability has a limited role in determining the type and amount of goods produced. Market forces, Coase postulated, govern production regardless of how the legal system assigns liability. While

[10] Coase, "The Problem of Social Cost," 1–8.

[11] Coase, "The Problem of Social Cost," 1–8; and Ellickson, *Order without Law*, 2–4.

Coase believed that laws should be crafted so as to reduce transaction costs, he argued that government intervention was neither as important nor as effective in shaping the economy as previously thought.[12]

Given the prominence of the Coase Theorem, it was perhaps inevitable that someone would conduct a study of how farmers and ranchers actually manage rampaging cattle. Robert Ellickson explained that as an expert in land-use law he had employed the Coase Theorem to explain how individuals bargain in a way that is to their mutual advantage. Born of a desire to venture beyond the walls of the law library, Ellickson went in search of and discovered in Shasta County, California, Coase's hypothetical ranchers and farmers in the flesh. Located near the Oregon border at the northern terminus of California's Central Valley, Shasta County is the home to farmers and ranchers living together in close proximity. In addition, land in the rural portions of Shasta County is designated as either of open and closed range. In open range, as defined in Shasta County, a rancher is not liable for damage inflicted by his herd unless the rancher intentionally engineers his livestock's trespass onto a neighbor's property or the animals breach a fence that is deemed legally sufficient. In all other circumstances, a rancher is without liability even if the trespass is the result of his negligence. A closed range, conversely, holds a rancher strictly liable for any damage caused by his herd. All of the pieces of the Coase Theorem were in place for an examination Ellickson likened to the anthropological inquiry of Clifford Geertz.[13]

Ellickson discovered that the farmers and ranchers of Shasta County do not behave in the manner suggested by the Coase Theorem. While the ranchers and farmers do cooperate in resolving problems arising from wayward cattle, they do so with a decidedly incomplete understanding of the substantive law or, even in some instances, whether the dispute took place on land designated open or closed range. The careful bargaining based on an understanding of legal rights and obligations that Coase imagined, simply did not take place. Instead, the residents of Shasta County operated pursuant to an unwritten code of behavior best characterized as neighborliness. A farmer, for example, is likely to suffer occasional invasions by cattle without complaint, mentally calculating any damage as a debt to be repaid by the rancher in some other situation. Neighbors in rural Shasta County deal repeatedly with one another and find ways to even up their accounts. The rancher, Ellickson found, might bear a greater burden of cost or work on a community water project. Deviants who do not abide by the community norms are punished primarily by self-help measures taken by the aggrieved. Gossip is generally an effective way of coercing the recalcitrant into compliance. In a community where families have lived for many generations and a good reputation for neighborliness is valued, spreading word of an unbalanced account is generally sufficient to secure that parity is

[12] Coase, "The Problem of Social Cost," 2–3.

[13] Ellickson, *Order without Law*, 1–9.

regained. Self-help may, on rare occasions, escalate to the point where a wayward animal is harmed in such a way that a clear message is sent to the offending rancher. It is only in the most egregious situations or those involving outsiders such as passing motorists that county officials are notified or litigation is commenced.[14]

While Coase correctly predicted that farmers and ranchers would manage to resolve potential disputes to their mutual benefit without recourse to governmental institutions, he failed, in Ellickson's estimation, to grasp the central insight to be gained from his Theorem. Government is not the sole provider of the rules which maintain order and permit society to operate efficiently. Neighbors, business associates, and all manner of people in everyday situations navigate their relationships according to principles and guidelines that supplement and, in some cases, even contradict state laws and regulations. Coase was unable to see the error in adopting the related fallacies of viewing the state as the source of all law – legal centralism – and the Hobbesian concept that government is the sole guarantor of social order. Ellickson further declared that people do behave rationally, but what constitutes the basis for rational action must be expanded beyond a narrow economic focus imposed by Coase.[15]

Building on the insight that Shasta County farmers and ranchers developed their own norms in large part because they perceived recourse to formal legal institutions to be extremely expensive and contrary to their ideal of neighborliness, Ellickson proposed that close-knit communities resolve disputes in ways that minimize transaction costs and maximize the welfare of the group. Ellickson, in *Order without Law* and an earlier law review article, offered as an example of this phenomenon the nineteenth-century American whaling industry. Whaling in the nineteenth century was, despite the disparate nationalities of participants and the global expanse of its fisheries, conducted by a close-knit community. As Ellickson explains, whalers met constantly at sea exchanging information. In addition, their "home and layover ports were few, intimate, and socially interlinked." Whaling was also an intensely competitive business and disputes over which of several pursuing ships was entitled to a slain whale were inevitable. Absent some rules governing these situations, violence at sea would ensue. The rules that ultimately emerged were the product of customs that evolved among the community of whalers. In order for these norms to be effective – in Ellickson's estimation – the involved parties had to understand exactly what sort of behavior was expected. English and American courts, recognizing the virtue of these participant-created norms, invariably deferred in Ellickson's estimation to what they perceived to be the custom of the industry in deciding the handful of whale capture disputes that resulted in litigation. Ellickson's story of nineteenth-century whaling has, in effect, a happy ending.

[14] Ellickson, *Order without Law*, 1–10.

[15] Ellickson, *Order without Law*, 4–6.

Whalemen maximized their welfare and provided the valuable lesson to society that allowing individuals to shape the laws by which they are governed brings more goods to market at a lower cost than when rules are imposed by outsiders.[16]

Ending his tale in the nineteenth century, Ellickson is able to elide the issue of resource depletion. Legal scholar R. Brent Walton has vigorously challenged Ellickson's theory of whaling's welfare maximization by pointing out that whale stocks were reduced throughout the nineteenth century by the very efficiency that made New Bedford an economic powerhouse by the 1830s. Walton's point is that the whaling industry committed suicide in exchange for short-term maximization of welfare and wealth. Ellickson offers two responses that illustrate his minimal concern with the connection between norms and resource depletion. Conceding that the overfishing of certain types of whales which led to longer voyages and ever more distant fisheries was not welfare maximizing, Ellickson observes "that norms that enrich one group's members may impoverish, to a greater extent, those outside the group." Norms, Ellickson also reflects, are ill equipped to either master the type of sophisticated understanding of cetacean habitats and breeding or to implement the world-wide network of monitoring required to forestall depletion of whale stocks. "For a technically difficult and administratively complicated task such as this," Ellickson concedes, "a hierarchical organization, such as a formal trade association or a legal system, would likely outperform the diffuse social forces that make norms."[17]

Ellickson's inability to provide a compelling response to Walton's challenge is a reflection of the degree to which he has essentialized whalemen into welfare-maximizing machines. They are welfare maximizing, Ellickson's circular logic suggests, because that is what they do. Whalemen certainly killed a large number of whales, but it is not really possible to state with any authority whether their practices maximized profits or were, for that matter, particularly efficient. When confronted with the inability or disinterest of whalemen in preserving whale stocks, Ellickson is left to concede that perhaps their ability to maximize their welfare was limited to short periods of time and small groups. While Ellickson offers a more nuanced approach to what motivates human action than Coase, his vision of whalemen as effective maximizers of their welfare forces him into the position that his protagonists agreed upon and followed the most efficient course of action in dividing disputed whales. This, in turn, prevents Ellickson from seeing the uncertainty and flexible negotiations that actually attended property conflicts at sea or in port. The reality of general

[16] Ellickson, *Order without Law*, 191–206, 193; and Robert C. Ellickson, "A Hypothesis of Wealth-Maximizing Norms: Evidence from the Whaling Industry," *Journal of Law, Economics, and Organization* 5 (Spring 1989): 83, 85. *Fennings v. Lord Grenville*, 1 Taunt. 241 (1808).

[17] Ellickson, *Order without Law*, 205, 206. R. Brent Walton, "Ellickson's Paradox: It's Suicide to Maximize Welfare," *New York University Environmental Law Journal* 7 (1999): 153.

principles subject to much dispute and confusion as to proper behavior is at odds with Ellickson's vision of universally accepted norms well suited to the type of whale hunted in a particular fishery.[18]

Attention to the issues of stock depletion raised by Walton in critique of Ellickson does not guarantee an understanding of whaling practices that is any less static and essentialized. Models of resource depletion offered by scientists and environmental historians have been no less prone to reducing human motivation to a single imperative driving all of their actions and decisions. Dubbed "the tragedy of the commons," Garrett Hardin's iconic description of the inevitable destruction of commonly owned and competitively hunted resources was driven by his belief that humans are motivated exclusively by the desire to maximize their individual profit. Hardin's dark vision captured the imagination of scientists, historians, and legal theorists, spawning a cottage industry of scholarly work dissecting and largely accepting the idea that – unless checked – resources not owned by individuals or subject to strict regulation are doomed to destruction.[19]

When Garrett Hardin delivered his 1968 presidential address to the Pacific Division of the American Association for the Advancement of Science, he identified his subject as the "population problem." Hardin argued that the finite resources of the world can support only a limited number of people. Invoking the specter of Malthus, Hardin warned that only zero population growth would prevent an otherwise inevitable disaster of hunger and desperation. Hardin assumed that his audience at Utah State University and the readers of *Science* – in which a reworked version of the speech was published as "The Tragedy of the Commons" – shared his belief in a coming population crisis. His main purpose was to convince the scientific community and the educated public that there was no technological solution to the problem. There was, in other words, no technique or invention that would permit the earth to sustain more than a particular population. Neither farming the sea nor any other scheme or invention practical in the foreseeable future would stop the coming disaster caused by unchecked population growth. Instead, the freedom to reproduce at will must be replaced by a system of controls that effectively curbs the human tendency to self-destruction.[20]

Recognizing that the existing laissez-faire approach to reproduction enjoyed a cultural legitimacy inherited from the economic theory of Adam Smith, Hardin vigorously attacked the idea that individuals acting in their own self-interest would make decisions that would prevent overpopulation. The illustration Hardin employed to dispute the notion that unrestrained individuals and markets can be counted on to act in a way that benefits society came to overshadow in intellectual and academic discourse the looming population

[18] Ellickson, *Order without Law*, 192–206.
[19] Garrett Hardin, "The Tragedy of the Commons," *Science* 162 (December 13, 1968): 1243–8.
[20] Hardin, "The Tragedy of the Commons," 1243.

problem it was designed to illuminate. Long after Hardin's ideas about population had been largely dismissed as the product of a misanthropic mind, the pasture open to all herders he asked his audience to imagine has endured as the controlling narrative of how people deplete natural resources. Hardin explained that what he termed the tragedy of the commons arises because it is in the economic interest of each herder to keep as many animals as possible on the pasture. The problem arises when the number of animals exceeds the carrying capacity of the land. An individual herder is faced with the decision of whether to add to his own herd even though that additional animal will lead to further overgrazing of the field. Hardin explained that while a particular herder will enjoy the economic benefit of each additional animal, the cost of overgrazing will be shared by all of the herders. A rational herder, seeking the maximum profit and recognizing that his forbearance will not prevent his cohorts from increasing their herd, will inevitably add to his own collection of ungulates. The end result is the ruin of the pasture and the financial fortunes of all of the herders. The land, like all valuable resources owned in common and competitively exploited, is destroyed because of the aggregate effect of individual – often well-meaning – decisions.[21]

Hardin offered several solutions to the tragedy of the commons as it applied to pastures and other resources. The easiest resolution is to turn the commons into private property. Individual owners, recognizing that their long term economic well being is dependent upon the continued productivity of a piece of land or resource, are far more likely to act as responsible stewards. Privatization is, of course, virtually impossible to achieve for resources harvested in certain environments such as at sea. The most extreme way out, assuming the possibility of enforcement, would be a complete prohibition on use of the resource. This would deny everyone the benefit of the resource and thus be unacceptable to most people. Hardin advocated a regime of temperance in resource use that is the product of mutual coercion that is mutually agreed upon by most of the affected population. Acknowledging that taxation or any other effective system of coercion leads inevitably to complaints of constricted freedom, Hardin countered that those "locked into the logic of the commons are free only to bring on universal ruin." Education as to the consequences of inaction is required to convince people of the necessity of mutual coercion. [22]

The tragedy of the commons was never really designed to explain the process by which resources are exhausted and, in some cases, face extinction. No scholar revealed its limitations in explaining environmental change more clearly than Arthur McEvoy. McEvoy's *The Fisherman's Problem: Ecology and the Law in the California Fisheries, 1850–1980* chronicles the historical interplay of the plant, animal, and human communities that share California's rivers and

[21] Hardin, "The Tragedy of the Commons," 1244.
[22] Hardin, "The Tragedy of the Commons," 1248.

offshore fisheries. What McEvoy sees as the wanton destruction of California's fisheries is the result of "the repeated failure of public agencies to take effective action against the depletion of resources that, given a modicum of care, might have remained productive indefinitely."[23] The disasters that have befallen the fisheries of California are not, in McEvoy's estimation, tragic in the way Hardin imagined. McEvoy challenges the immutability of Hardin's tragedy of the commons with the observation that for Hardin "*Homo sapiens* was *Homo economicus*, a rational, individual, and wealth-maximizing creature."[24]

Concerned with an understanding of how ecology and law fit together and not so much with grand theoretical formulations, McEvoy counters that there is no such thing as human nature that governs the behavior of all people at all times. How people behave is the product of their culture. Accordingly, Hardin's herders "are not tragic in the sense that their undoing flows from some flaw in their inherent nature; rather, they are products of a particular culture with a particular history and a particular view of the world." The tragedy of the commons is, thus, not an inevitable outcome when commonly held resources are competitively sought. Instead, resource depletion springs, in part, from a particular cultural outlook that governs a society's use of resources and their perception of and response to resource problems.[25]

In addition to historicizing the commons, McEvoy insisted that fish and the water in which they swim must be considered in any explanation of the California fisheries. McEvoy explained that in the early 1950s a group of reformers, whose scholarship Hardin employed and ultimately recast as the tragedy of the commons, sought to remedy flaws in resource governance that they held responsible for such disasters as the post–World War II collapse of sardine stocks. What Hardin and his colleagues challenged was not the questionable scientific basis for calculations of what a fishery could safely yield, but the efficacy of any limit in a legal system that permitted renewable resources to be treated as common property. The tragedy of the commons model was born of the inability of government to enforce limits and the failure of market forces to restrain fishermen and harvesters of other products from endangering their quarry. The solution was to place endangered resources under the control of a single person or entity. This is what Hardin had in mind when he suggested that the way out of the commons dilemma could be found in privatization or the creation of coercive resource allocation schemes agreed to by industry participants. Under private or communal control, market incentives would effectively encourage harvesters to limit their take.[26]

[23] McEvoy, *The Fisherman's Problem*, xi.
[24] McEvoy, *The Fisherman's Problem*, 13.
[25] McEvoy, *The Fisherman's Problem*, 14.
[26] Arthur F. McEvoy, "Toward an Interactive Theory of Nature and Culture: Ecology, Production, and Cognition in the California Fishing Industry," in *The Ends of the Earth: Perspectives on Modern Environmental History*, ed. Donald Worster (Cambridge: Cambridge University Press, 1988), 211–29; McEvoy, *The Fisherman's Problem*, 10–14.

McEvoy is not the only scholar who has pointed out the ahistorical nature of Hardin's model. E. P. Thompson also recognized that Hardin's Malthusian vision of commonly held resources as doomed to ruin by some sort of "inexorable economic logic" denies the demonstrated ability of some communities to regulate their affairs in ways that do not lead to depletion and destruction of their means of making a living. While the Marxist Thompson and Robert Ellickson – a proponent of the law and economics school of legal thought which questions the wisdom of excessive government regulation of commerce – have little in common politically, they share with McEvoy the belief that close-knit groups can establish norms and customs that work to the benefit of the entire community.[27]

Anglo-American whalemen in the eighteenth and nineteenth centuries certainly shared the goal of maximizing their catch. They also – like most businessmen – were not anxious to come to blows or waste time and money litigating disputes. Shared goals did not, however, translate into universally accepted norms that in their representation of the community maximized profits and kept disputes at a minimum. While much of what Ellickson argues about the ability of close-knit groups to resolve disputes without recourse to the formal institutions of the law is accurate, he is mistaken in seeing that as the product of community-created customs and norms that were recognized by all participants. Instead, whalers shared a general sense that effort and persistence in pursuing a whale mattered in determining ownership. Vague and difficult to articulate, whaling customs were not consciously created by whalemen to serve shared goals. They evolved and were accepted because they worked. Like all evolutionary processes, the formation of whaling practices was complicated. When courts were called upon to pass judgment on the property law of whaling, they – like most scholars – were anxious to impose an order and consistency that did not exist at sea.

[27] E. P. Thompson, *Customs in Common* (New York: The New Press, 1991), 107. Thompson's analysis of Hardin is discussed in Ted Steinberg, "Down to Earth: Nature, Agency, and Power in History," *American Historical Review* 107 (June 2002): 818. McEvoy, "Toward an Interactive Theory," 227.

2

The biology, economics, and geography of Anglo-American whaling

Whaling in the eighteenth and nineteenth centuries was an important global industry. Ships from many nations with crews from ports all over the world hunted in waters from the Arctic Ocean to the Tasman Sea. Whale oil illuminated the cities and greased the machines of the Industrial Revolution. While sperm and bowhead whales – among the principal targets of commercial whaling in this period – were relatively plentiful by current standards, they were always scarce in the sense that they were difficult to locate, chase, and kill. A profitable season of whaling could be gained by the rendering of less than ten whales. In an extractive industry where a ship could go weeks without even sighting a whale, each encounter with a possible catch was fraught with tension and the recognition that the success of a season or even a multi-year voyage depended upon a handful of such attempts to land their prey.

Mastery of the taxonomy of the creatures known as whales is a daunting task. Whales, scientifically speaking, are members of "the phylum Chordata, the subphylum Vertebrata, the class Mammalia, and the order Cetacea." While a recitation of phylum, subphylum, class, and order bespeaks a high degree of precision, the order Cetacea encompasses about eighty-six species of a dizzying variety of sizes and shapes. Whales range in weight from that of the dwarf sperm at around 350 pounds to the 300,000 pound blue whale. The most useful distinction between species of whales for an understanding of whaling is in the cetacean suborders of Mysticeti (baleen whales) and Odontoceti (toothed whales). The appellation of the latter is somewhat misleading as not all of the over seventy species of Odontoceti have functional teeth and the suborder includes dolphins and porpoises.[1]

[1] Lance E. Davis, Robert E. Gallman, and Karin Gleiter, *The Pursuit of Leviathan: Technology, Institutions, Productivity, and Profits in American Whaling, 1816–1906* (Chicago: University of Chicago Press, 1997), 20–1, 52–5; Karl Brandt, *Whale Oil: An Economic Analysis* (Palo Alto:

FIGURE 1 Greenland whale; Sperm whale. Naturgeschichte der Säugethiere: mit colorirten Abbildungen zum Anschauugs-Unterricht für die Jugend, 1872.
Picture Collection. The New York Public Library, Astor, Lenox, and Tilden Foundations

Sperm whales, the only Odontoceti widely pursued by commercial hunters prior to 1880, range in maximum size from forty feet and eighteen tons for females to sixty feet and fifty tons for males. Despite the two rows of large teeth arrayed along a sperm whale's narrow bottom jaw, their upper teeth are vestigial and their mouth configuration precludes effective chewing. Able to dive deeper than a mile and remain under the surface for as long as an hour in search of their preferred diet of small squid, the means by which sperm whales manage to catch their elusive prey remains unknown. Speculation ranges from the ability of sperm whales to emit loud noises which stun nearby squid or other small fish to the simple practice of swimming with its massive mouth open and then clamping down on any unfortunate visitors to its gaping maw. While

Stanford University Press, 1940), 11; Jeremy Cherfas, *The Hunting of the Whale: A Tragedy That Must End* (London: The Bodley Head, 1988), 15; Randall Reeves, "Cetacea (Whales, Dolphins, and Porpoises)," in *Grzimek's Animal Life Encyclopedia*, Michael Hutchins, Arthur V. Evans, Jerome A. Jackson, Devra G. Kleinman, James B. Murphy, Dennis A. Thoney, et al., eds., vol. 15: *Mammals IV*, 2nd edition (Detroit: Gale, 2004), 1–11; and Randall R. Reeves and Tim D. Smith, "A Taxonomy of World Whaling: Operations, Eras, and Sources," Northeast Fisheries Science Center Reference Document 03–12, www.nefsc.noaa.gov/publications/crd/crd0312/crd0312.pdf, accessed December 1, 2006. For a general introduction to sperm whale biology and distribution, see Simona L. Perry, Douglass P. DeMaster, and Gregory K. Silber, "The Sperm Whale," *Marine Fisheries Review* 61, issue 1 (1999): 59–74.

sperm whales can be found in water deeper than 300 meters around the world, their preference is for temperate and tropical regions. The congregations of female and immature sperm whales sometimes numbering in the hundreds which were so appealing to hunters rarely move beyond latitudes of 50 degrees north or south. Adult males venture into latitudes as high as 70 degrees in the summer, but their generally solitary nature at this time of year rendered them poor targets for commercial whaling. In the winter, sexually mature males answer the call of reproductive necessity and join their prospective mates in the vicinity of the equator.[2]

Female sperm whales commence ovulation at about the age of nine. With a calving interval of four to six years and the cessation of ovulation at some point in their forties, sperm whale communities – in the absence of significant immigration from other populations – do not grow rapidly. In addition to human predators, sperm whales are attacked by killer whales and also succumb to a number of viruses. As with all cetaceans, it is difficult to measure the life expectancy of sperm whales. Estimates in the range of sixty to seventy years may well understate their longevity by many years. At the beginning of extensive commercial whaling, the worldwide population of sperm whales – not counting the small numbers that frequented the North Atlantic – has been estimated in the range of 1,800,000 to 2,400,000. Unlike other cetacean populations, their wide range and near continuous occupation of the ocean renders difficult the study of discrete sperm whale communities. The degree to which human predation reduced sperm whale populations during the nineteenth century is the subject of much debate. Estimates range from as little as 30% to as much as 75%. The commercial hunting of sperm whales in the second half of the twentieth century was extensive and, according to those who place the nineteenth-century toll on the lower side, even more devastating than that previously inflicted.[3]

The distinctive appearance of sperm whales is the result of their massive heads with eyes set behind the corners of its mouth which curl up in a sort of smile. Much of a sperm whale's head which accounts for roughly one-third of the creature's length contains the valuable spermaceti sought by generations of whalemen. The upper portion of a sperm whale's head holds the conical shaped spermaceti organ from which twenty or so barrels of nearly pure spermaceti oil can be scooped. Below the spermaceti organ is a mass of fibrous material – termed junk by whalemen – containing more of the valuable oil that, along with the ligaments beneath and the blubber from the rest of the whale,

[2] Perry, DeMaster, and Silber, "The Sperm Whale," 59–74; and Eric Jay Dolin, *Leviathan: The History of Whaling in America* (New York: W. W. Norton & Co., 2007), 75–84.

[3] Perry, DeMaster, and Silber, "The Sperm Whale," 59–74; Davis, Gallman, and Gleiter, *The Pursuit of Leviathan*, 132–3; Hal Whitehead, *Sperm Whales: Social Evolution in the Water* (Chicago: University of Chicago Press, 2003), 11, 115, 122–6. ; and Tim Smith and Randall Reeves, "Design of a Program on Sperm Whale Catch History: Results of a Workshop," at www.worldwhalinghistory.org/DesignofaProgramofResearchonSpermWha.pdf, accessed October 10, 2009.

can be boiled to produce more oil. The largest sperm whales produced upward of forty barrels of oil. The exact purpose of this large reservoir of spermaceti has long eluded scientists, but it appears that it plays a role in the loud clicks produced by sperm whales as an aid in echolocation and communication.[4]

In the first decades of the nineteenth century, sperm oil was used primarily for illumination. Sperm oil, burning more brightly than the oil of other cetaceans, was the favored fuel for lighthouses and city streets. Home use of the odorless and clean burning sperm oil and candles made from pure spermaceti was limited by cost to only the wealthiest consumers. Sperm oil was generally two to three times as expensive per gallon as the oil of baleen whales. By the second quarter of the nineteenth century, sperm oil came increasingly to be used as an industrial lubricant in the manufacture of cotton and woolen goods. Growing domestic competition from gas and other sources of illumination in combination with rising demand from Great Britain during the 1840s effectively reduced the American market of sperm oil to industrial lubrication. More than half of the sperm oil produced in America was consumed by the British in the years just prior to the Civil War. From a high of nearly five million barrels produced in the period from 1840 to 1844, the American output of sperm oil fell to about 1,300,000 barrels for 1865 to 1869.[5]

The reputation of sperm whales as deadly adversaries for the men who dared heave harpoons from small boats is well deserved. Sperm whales possess strong tails that on many occasions were used to smash the small boats dispatched from whaling ships. Using its massive head as a sort of demolition ball, a sperm whale could also destroy a boat or seriously damage a ship. It was, however, the sperm whale's lower jaw that particularly terrified whalemen. Sperm whales will roll onto their backs, open their mouths, and expose their teeth just above the surface of the water. While sperm whales only rarely snapped their jaws shut onto a boat or crew, the prospect of such an attack was often enough to convince men when so confronted to leap into the water. Bowheads and right whales – the primary eighteenth- and nineteenth-century Mysticeti targets of whalemen – have neither the quickness nor the saw-like jaw of the sperm whale, but they also presented significant dangers to their hunters. Often characterized as docile in comparison with sperm whales, bowheads and right whales possess tails and heads capable of much destruction. Yet, it was the icy environment of the arctic and subarctic waters inhabited by bowheads that does much to explain why their pursuit was every bit as dangerous as the hunt for sperm whales. Dragged through fields of ice in dense fog, whalemen were regularly required to quickly cut attached lines or be carried to their death by a sounding bowhead. Even without confronting a bowhead, sheets of ice driven by rapidly

[4] Dolin, *Leviathan*, 75–84; and Whitehead, *Sperm Whales*, 8–11.

[5] Davis, Gallman, and Gleiter, *The Pursuit of Leviathan*, 342–68.

shifting winds could splinter vessels or strand a ship until the coming of the spring thaw.[6]

While sperm whales range across the great expanse of the world's oceans, the much smaller population of bowheads occupied limited areas dictated by their particular food and temperature requirements. Biologists have identified five different historic stocks of bowheads: the Spitzbergen, Davis Strait, Hudson Bay, Bering-Chukchi-Beaufort, and Okhotsk Sea stocks. The Spitzbergen stock off the east coast of Greenland – originally numbering about 25,000 bowheads – was the target of the first sustained pelagic whaling in the early seventeenth century. By the 1820s, the Dutch and the English had driven the population of the Spitzbergen and 11,000 or so Davis Strait bowheads to the brink of extinction. The Hudson Bay stock which numbered only about 600 whales prior to the beginning of commercial exploitation was quickly depleted in the 1860s.[7]

American whalers entered the South Pacific in the 1790s and began hunting the 18,000 bowheads of the Bering-Chukchi-Beaufort stock in the 1840s. In 1848, the *Superior* of Sag Harbor successfully navigated the Bering Strait and opened the Chukchi and Beaufort Seas to American vessels in search of bowheads. By 1850, the Sea of Okhotsk was opened to bowhead whaling. While scientists are divided as to whether the Atlantic and Bering-Chukchi-Beaufort bowhead stocks are reproductively separate, it is generally agreed that the Okhotsk Sea stock is entirely isolated from the other populations. Numbering only about 3000 to 6500 at the onset of commercial whaling, the bowheads of the Sea of Okhotsk were severely depleted by 1870. The small number of bowheads prior to commercial whaling, their limited geographic range, the female's capability of birthing a calf only every three to four years, late sexual maturation – fifteen to twenty years – and a gestation period in excess of thirteen months worked in combination to limit their ability to survive a determined human predator.[8]

[6] Dolin, *Leviathan*, 88–9. For a discussion of the hazards of hunting bowheads in icy waters, see William Scoresby, *An Account of the Arctic Regions With a History and Description of the Northern Whale-Fishery*, 2 volumes, (Edinburgh: Archibald Constable and Co., 1820), II: 346–68.

[7] Douglas A. Woodby and Daniel B. Botkin, "Stock Sizes Prior to Commercial Whaling," in *The Bowhead Whale*, ed. John J. Burns, J. Jerome Montague, and Cleveland J. Cowles (Lawrence, KS: The Society for Marine Mammalogy, 1993), 387–407; Kim E. W. Shelden and David J. Rugh, "The Bowhead Whale, Balena Mysticetus: Its Historic and Current Status," *Marine Fisheries Review* 57, issue 3/4 (1995): 1–21; Robert C. Allen, "The First Great Whale Extinction: The End of the Bowhead Whale in the Eastern Arctic," *Explorations in Economic History* 38 (2001): 448–77; Phillip J. Clapham, Sharon B. Young, and Robert L. Brownell, Jr., "Baleen Whales: Conservation Issues and the Status of the Most Endangered Populations," *Mammal Review* 29, no. 1 (1999): 35–60; Jon M. Conrad, "Bioeconomics and the Bowhead Whale," *The Journal of Political Economy* 97, no. 4 (August 1989): 974–87; and Robert C. Allen and Ian Keay, "Saving the Whales: Lessons from the Extinction of the Eastern Arctic Bowhead," *Journal of Economic History* 64, issue 2 (June 2004): 400–32.

[8] Wodby and Botkin, "Stock Sizes Prior to Commercial Whaling," 387–407; Shelden and Rugh, "The Bowhead Whale," 1–21; and John R. Bockstoce, *Whales, Ice, and Men: The History of Whaling in the Western Arctic* (Seattle: University of Washington Press, 1986).

FIGURE 2 Bowhead whale head on deck.
Courtesy of the New Bedford Whaling Museum

FIGURE 3 Baleen in warehouse of William Lewis and Son.
Courtesy of the New Bedford Whaling Museum

Baleen whales, including bowheads, possess neither the teeth nor the oral construction to consume the squid favored by sperm whales. Mysticeti, instead, feed by taking in large amounts of seawater and using the two racks of baleen attached to their palate to strain out the small creatures that provide their nourishment. The size and structure of baleen, or whalebone as it was generally known, differs between species of Mysticeti. An adult bowhead possess about 250 to 350 individual baleen plates per side which, depending upon their location along the extremely arched upper jaw, can be up to 4.6 m long. Individual plates are of a thin, elongated triangular shape that project from the maxilla parallel to one another at a distance of about a centimeter. Baleen, like horns and hoofs, consists of hard keratins that with the application of heat can be bent into new shapes that are retained when cooled. This attribute of whalebone, in combination with its flexibility, accounted for its commercial value in objects from corsets to umbrellas. The hard interior edge of the baleen is degraded by the bowhead's massive tongue revealing a cemented hairlike substructure that as it frays creates the sieve with which the whale strains its food.[9]

At an average body length 12 to 18 m, the caloric requirements of bowheads are enormous. One study suggests that even a bowhead of only 13.7 m would need to annually ingest 4000 kg of lipids. Digesting primarily copepods, euphausiids, and other types of zooplankton less than 30 mm long, bowheads must consume about 100 mt of crustaceans during an approximately six month feeding season. It is estimated that the daily intake during feeding season ranges from about 400 to 2100 kg. To meet their prodigious nutritional requirements, bowheads continuously swim slowly with an open mouth through fields of prey in a process termed ram feeding. The structure of a bowhead's mouth and skull is well adapted for taking in an enormous amount of water, straining out food, expelling the water, and using its tongue to move feed toward the esophagus. It is speculated that the morphology of the bowhead feeding apparatus creates a hydrodynamic pressure that facilitates the entry of water into the whale's mouth without alerting prey to their impending demise. While the particulars of the bowhead diet and feeding habits are not well understood, scientists agree that in order to survive bowheads must spend at least half their time in water with an extraordinarily high biomass of prey. In addition to the exigencies of securing sufficient nutrition, bowhead range is also restricted to icy subarctic

[9] Lloyd F. Lowery, "Foods and Feeding Ecology," in Burns, Montague, and Cowles, 201–38; Alexander J. Werth, "Models of Hydrodynamic Flow in the Bowhead Whale Filter Feeding Apparatus," *The Journal of Experimental Biology* 207 (2004):3569–80; Richard H. Lambertsen, Kerry J. Rasmussen, Winston C. Lancaster, and Ramond J. Hintz, "Functional Morphology of the Mouth of the Bowhead Whale and Its Implications for Conservation," *Journal of Mammalogy* 86, no. 2 (April 2005): 342–52; and Julie A. Lauffenburger, "Baleen in Museum Collections: Its Sources, Uses, and Identification," *Journal of the American Institute for Conservation* 32, no. 3 (1993): 213–30.

and arctic waters. With a layer of blubber of up to 43 cm, bowheads are equipped to live in temperatures that leaner Mysticetes cannot tolerate.[10]

The oil of bowheads and other baleen whales was known as whale oil and, like sperm oil, was used for illumination and industrial lubrication. Whale oil was generally more plentiful and almost always less expensive than sperm oil. In the period from 1830 to 1860, the American whaling industry produced approximately 387,000,000 barrels of whale oil and only about 221,000,000 barrels of sperm oil. Only in the years prior to 1830 and in the industry's final years after 1880 was more sperm oil than whale oil harvested by American whalemen. In 1814, the real value of unrefined whale oil at New Bedford was 0.77 dollars per gallon. This marked the highest price for whale oil until it was equaled in 1857 and the only year in the nineteenth century it fetched more on the New Bedford market than sperm oil. During most of the intervening years, whale oil was anywhere from 0.25 to 0.58 dollars per gallon. Whale oil reached its historic high in 1865 at 0.78 dollars per gallon and thereafter only surpassed 0.60 in three seasons prior to 1900. In 1840, the real price of unrefined sperm oil surpassed one dollar per gallon and remained above that mark in all but a few seasons until 1879 when it began a steady decline in value. During much of the nineteenth century, sperm oil cost two to three times the price of whale oil. Sperm oil's higher price was a function not only of supply, but its superior quality as both an illuminant and a lubricant. Whale oil produced a duller light than its competitor and emitted an unpleasant odor. As a lubricant, sperm oil – which remained liquid at lower temperatures than whale oil – was better suited than its competitor for the growing demands of lighter, faster machinery.[11]

Whalebone, exclusively the product of baleen whales, was until the end of the eighteenth century of little or no value and was generally discarded. While demand for the strong, flexible whalebone that was used in multiple products including whips and suspenders grew slowly over the first few decades of the nineteenth century, a change in women's fashion ignited the baleen market by the late 1830s. The fuller skirts of the 1840s and the hoop skirts of the 1850s and 1860s achieved much of their great expanse through the use of whalebone. The amount of baleen produced by American whalemen surged during the 1830s from 279,000 pounds in 1831 to 2 million pounds at the end of the decade. In 1853, production reached its zenith with nearly 5,700,000 pounds of baleen. The real price of unprocessed whalebone for the New Bedford market reflected this spike in demand. From 0.09 dollars per pound in 1820 to 0.16 in 1839, whalebone reached 0.99 dollars in 1858. The amount of whalebone

[10] Jerrold T. Haldiman and Raymond J. Tarpley, "Anatomy and Physiology," in Burns, Montague, and Cowles, 71–156; Lowery, "Foods and Feeding Ecology," 201–38; Werth, "Models of Hydrodynamic Flow," 3569–80; Lambertsen, Rasmussen, Lancaster, and Hintz, "Functional Morphology of the Mouth of the Bowhead Whale," 342–52; and Lauffenburger, "Baleen in Museum Collections," 213–30.

[11] Davis, Gallman, and Gleiter, *The Pursuit of Leviathan*, 342–68.

prepared for sale, thereafter, declined rapidly. By 1858, less than 2 million pounds was processed and 1867 marked the last season in which American whalemen brought to market in excess of 1 million pounds of bone. The price of whalebone, responding in part to the decline in demand, fell during the course of the 1860s. The meteoric rise in prices during the final quarter of the nineteenth century was once again ignited by women's fashion. Fulfilling the desire for a preternaturally small waist required the use of corsets for which the preferred material was baleen. The price of whalebone soared above 2 dollars per pounds in 1877 and peaked at just over 7 dollars in 1892. Had whalemen remained able to deliver the volume of bone at mid-century levels, prices would not have risen so dramatically by the 1890s. Bowheads and right whales had become harder to find and with the market for oil severely restricted, whalemen relied on whalebone to keep voyages financially viable. Whalebone, which began the nineteenth century as an afterthought, was, by 1900 the commodity whalemen prized. The once valuable blubber was now often left to rot.[12]

As economic historians Lance E. Davis, Robert E. Gallman, and Karin Gleiter have noted in their exhaustive study of American whaling in the nineteenth century, calculating profits for any business is extremely difficult "both theoretically and empirically." Davis, Gallman, and Gleiter have determined whaling profits for individual voyages by subtracting costs from the real value of the returned whale products. While determining the real value of the catch – outputs of oil and bone multiplied by average prices in the year the vessel returned and then divided by a general price index – is relatively simple, calculating costs is much more speculative. They included among the costs the following factors: cost of crew subsistence, the crew's share of output, the value of items such as sails and rigging consumed, depreciation, agent fees, and insurance. Comparing the profits generated by American voyages to the Western Arctic and the Pacific Ocean at latitudes above 50 degrees north with estimates of the number of whales taken per ship in those waters reveals just how important a single whale lost or taken could be in whether a vessel enjoyed a successful season.[13]

[12] Driven in large part by the demands of the fashion industry, much of the whalebone gathered by American ships throughout the nineteenth century was sent to France and other countries in northern Europe. In the 1850s, approximately 70% of the American harvest of whalebone was exported. Davis, Gallman, and Gleiter, *The Pursuit of Leviathan*, 342–68; Bockstoce, *Whales, Ice, and Men*, 29, 208, 220; and Dolin, *Leviathan*, 356–62.

[13] Davis, Gallman, and Gleiter, *The Pursuit of Leviathan*, 423–58, 423. For statistics from the Western Arctic and Pacific fisheries, see Alexander Starbuck, *History of the American Whale Fishery From its Earliest Inception to the Year 1876* (New York: Argosy-Antiquarian Ltd., 1964), 2 volumes, I: 104. Rather than receive salaries, American whaling captains and crews were each paid a fraction of the net catch. The percentage or lay paid to an individual was a reflection of his perceived value. Captains typically received about a 1/15th share. The least

The annual average number of baleen whales – mostly bowheads – taken by American vessels in the Western Arctic and the North Pacific after the opening of these fisheries in the late 1840s fluctuated from about 7 to 15. During the years from 1853 to 1857 the annual average take of the approximately 200 ships in the American fleet was 12.3 whales. Davis, Gallman, and Gleiter calculate the mean annual average profits by percentage for this period to be 22.6. Over the next four years the average take of whales by a fleet of approximately 150 ships each season dropped to 8.7. With the declining catch came a reduction of profits to 12.3% for a fall of about 45%. The connection between the number of whales taken in a year and seasonal profits is far from direct. In particular years, the price of baleen and oil might be high enough to rescue a season that would otherwise have been deemed a bottom line failure. With a seasonal average of only 11.2 whales for the period from 1851 to 1870, the rendering of each cetacean represented a rather significant portion of the hoped for income of crews and owners. This basic economic reality of whaling was never lost on whalemen, but was never really grasped by the lawyers and judges in the disputes over the possession of whales that reached British, American, and Hawaiian courts. Whaling, in the eyes of the law, was much like hunting a rabbit.[14]

Whalemen, of course, knew better. Hunting any species of whale in the nineteenth century was a dangerous business. While whaling would later take on the impersonal qualities of an industrialized enterprise, it was for earlier generations of whalemen an intensely up-close and personal battle with a powerful animal. Once American whaling moved by the eighteenth century from coasts and bays and out into open seas, the basic method of hunting was set and the industry resisted significant changes in technique or equipment throughout the nineteenth century. Whales were pursued by crews of six or so men in small boats dispatched from ships. Ships typically had five shallow, open boats of up to thirty feet in length and six feet in width ready to be lowered into the water at any time a target was spotted or, in the ice of the arctic, when conditions prevented a ship's passage to where whales were likely to be found. The ideal whaling boat was both light and strong. Crews often had to row significant distances in pursuit of whales and then withstand the forces of a fleeing animal desperate to escape the grasp of an attached harpoon and rope.[15]

experienced member of a crew might earn only 1/200th of net value of the take. Davis, Gallman, and Gleiter, *The Pursuit of Leviathan*, 150–202.

14 Davis, Gallman, and Gleiter, *The Pursuit of Leviathan*, 423–58; Starbuck, *History of the American Whale Fishery*, I: 104; and Box 3, Part 1, #17, David A. Henderson Collection, New Bedford Whaling Museum, New Bedford, Massachusetts (NBWM) #2008.18.

15 Davis, Gallman, and Gleiter, *The Pursuit of Leviathan*, 283–94; Bockstoce, *Whales, Ice, and Men*, 61–9; and Dolin, *Leviathan*, 49–50, 248–50. For general information on the equipment used to kill whales, see James Temple Brown, *The Whale Fishery and Its Appliances* (Washington: Government Printing Office, 1883).

The equipment used by whalemen necessitated that the combat be at close range. Whaling boats were positioned in contact with or close enough to their target to permit the harpooner or, as he was more commonly called, the boatsteerer to throw a harpoon with enough force to pierce the skin of the targeted whale. The harpoons or irons, in the industry's argot, were attached to the trunk of a very small tree which allowed the harpooner a surer grip for a more accurate and vigorous thrust. The most adept boatsteerers were able to strike a whale with sufficient force from about thirty feet. Designed to prevent the harpoon from drawing loose during a protracted chase, harpoon tips were fashioned with a single or double barb or by the middle of the nineteenth century a toggle.

The harpoon did little physical harm to the whale but did cause the hasty retreat which might last for hours before the crew could once again draw up

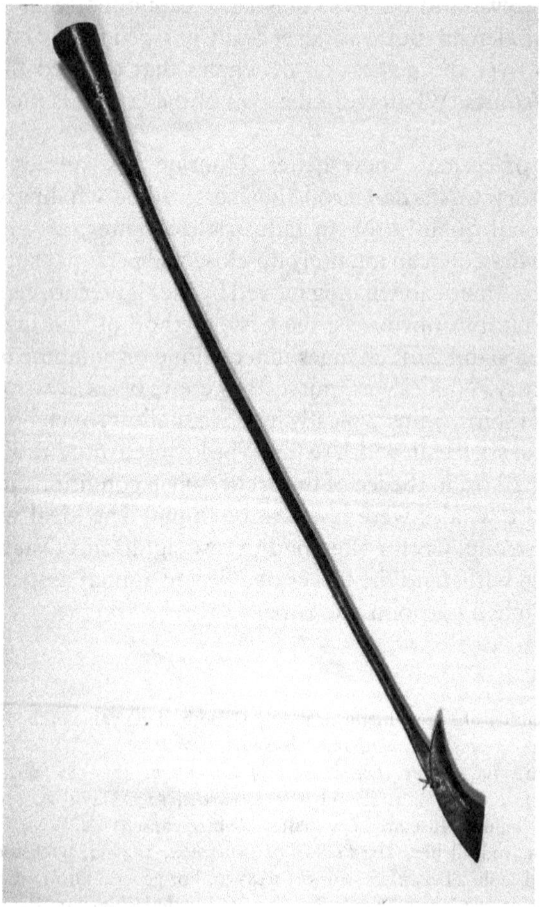

FIGURE 4 Toggle iron.
Courtesy of the New Bedford Whaling Museum

close enough to the now exhausted whale and inflict the ultimately fatal blows. A handheld spear called a lance was then used to repeatedly stab the whale until it succumbed.[16]

For well over a hundred years, American whalemen hunted their prey in much the same way regardless of the fishery or the species pursued. The speed and ferocity of sperm whales and their tendency to travel in pods encouraged the use of pieces of wood called drogues which were attached to the harpoon lines and served much the same purpose as an attached boat in slowing down and tiring a fleeing whale. Use of drogues permitted a whaling crew to attack – at least in theory – multiple sperm whales simultaneously, but in reality a single boat was generally forced to chase only one whale at a time. The icy waters frequented by bowheads, however, also made drogues useful when whalemen were forced to sever the connection between the target and the boat. A crew remained attached to a diving whale in a field of ice at their own peril. While chasing bowheads and sperm whales presented different challenges the same basic procedures and techniques prevailed.[17]

Nineteenth-century American whalemen were generally resistant to change. They usually eschewed the corporate structure for financing voyages and were slow to abandon wind and sails long after other maritime industries adopted steam power. Whalemen also displayed a marked preference for human muscle over explosives in fastening to whales. By the early eighteenth century a rocket powered harpoon was developed but was little used as it was difficult to aim in rough waters and had a significant kick that injured men and damaged boats. Even after improved weapons for firing harpoons were commercially available by the 1840s, American whalemen continued to throw harpoons at whales from close range. Whaling crews were not – at least in this regard – just narrow-minded traditionalists. They simply did not believe that the new methods for propelling harpoons represented any great improvement over the established way of doing things. This new technology was only adopted in that very limited segment of the American whaling industry which targeted rorquals such as blue and fin whales which were such fast swimmers that drawing up within throwing range proved difficult.[18]

Explosives proved, however, much more useful in killing whales that had been harpooned. Invented in the 1840s and in general use by the middle of the next decade, the so-called bomb lance was shot from a gun and delivered

[16] Davis, Gallman, and Gleiter, *The Pursuit of Leviathan*, 283–94; Bockstoce, *Whales, Ice, and Men*, 61–9; Brown, *The Whale Fishery and Its Appliances*, 6–9, 12–4; and Dolin, *Leviathan*, 49–50, 248–50.

[17] Davis, Gallman, and Gleiter, *The Pursuit of Leviathan*, 283–94; Bockstoce, *Whales, Ice, and Men*, 61–9; Brown, *The Whale Fishery and Its Appliances*, 33; and Dolin, *Leviathan*, 49–50, 248–50.

[18] Davis, Gallman, and Gleiter, *The Pursuit of Leviathan*, 283–94; Bockstoce, *Whales, Ice, and Men*, 61–9; Brown, *The Whale Fishery and Its Appliances*, 9–12; and Dolin, *Leviathan*, 49–50, 248–50.

FIGURE 5 “Trying Out.” Harper’s Magazine, 1874.
Picture Collection. The New York Public Library, Astor, Lenox, and Tilden Foundations

an explosive charge seconds after it pierced the whale’s hide. Bomb lances were particularly useful in the bowhead fisheries as they offered the possibility of killing a whale before it dove under the cover of ice. Sperm whalers did not face icy conditions and believed that the sound of the guns frightened and dispersed their prey which tended to swim in pods. While effective and widely

FIGURE 6 Tryworks at full blast.
Courtesy of the New Bedford Whaling Museum

recognized in bowhead fisheries as an improvement in the tedious and often dangerous task of repeatedly thrusting a conventional lance into the body of a displeased cetacean, bomb lances were not without problems. They misfired, on occasion, injuring crew members. More frequently, they blew the attached harpoon and line out of the whale allowing the injured prey to escape. A type of human powered harpoon fitted with a gun that delivered an explosive charge into a whale was invented in the 1860s and quickly became popular in arctic waters. Known as a darting gun, this device was certainly safer than the bomb lance but it was heavy and difficult to throw. It was only effective in the hands of a boatsteerer in very close proximity to the targeted whale.[19]

Having killed a whale, crews faced the tedious task of towing it back to the ship for processing. Once along the ship's starboard side, the carcass was chained to the vessel. A stage was thereafter lowered around the whale and the process called "cutting in" commenced. The whale's head was removed first and then large blankets of blubber were carved out of the whale and hauled on deck. Since the eighteenth century, American whalers were equipped with what were called trypots for the boiling of blubber into oil which was then placed in wooden casks and stored below deck.

[19] Davis, Gallman, and Gleiter, *The Pursuit of Leviathan*, 283–94; Bockstoce, *Whales, Ice, and Men*, 61–9; Brown, *The Whale Fishery and Its Appliances*, 14–6; and Dolin, *Leviathan*, 49–50, 248–50.

The heads of whales required special attention. Baleen was cut from the mouths of bowheads and spermaceti was removed from a compartment in the heads of sperm whales. The entire process was messy, tedious, and dangerous. Men wielded extremely sharp blades from precarious positions and lifted great weights as the ship rolled and bobbed in water often filled with sharks drawn by the prospects of a grand feast.[20]

[20] Davis, Gallman, and Gleiter, *The Pursuit of Leviathan*, 273–78; Bockstoce, *Whales, Ice, and Men*, 74–92; Brown, *The Whale Fishery and Its Appliances*, 21–22; and Dolin, *Leviathan*, 265–69.

3

Fast-fish, loose-fish

The story of how British whalemen resolved disputes in the Greenland fishery and the way in which English and Scottish courts adjudicated the handful of litigated cases makes clear, whaling law was not created entirely at sea. British lawyers, judges, and legal scholars of the eighteenth and nineteenth centuries did not merely provide their imprimatur to that which was presented to them as whaling custom. The process by which the property law of whaling developed was, instead, collaborative. American judges and lawyers used these British cases when disputes were litigated in Boston and Honolulu because there were no applicable reported American cases. The British cases gave American jurists the mistaken impression that whalemen always employed well understood rules. While the British whaling norm of fast-fish, loose-fish was relatively well settled and generally followed in the Greenland fishery after the 1780s, American whaling customs grew increasingly muddled after an early preference for iron holds the whale weakened as the nineteenth century progressed.

I

January 1812 brought a familiar ritual to the port of Whitby on England's northeast coast. Captains began stocking their vessels for the summer season in the Greenland whalefishery.

Since 1753, Whitby had sent its ships into Arctic waters in pursuit of bowheads or what locals called the Greenland or Common whale. Captain William Scoresby, Jr. had every reason to be optimistic about the upcoming season as the meat of 26½ pigs was salted, packed, and stowed aboard the *Resolution* in early January.

The Yorkshire-born child of a whaling master of the same name, Scoresby was introduced to Arctic waters at age ten on his father's 1800 summer voyage to the Greenland fishery. By 1803, the younger Scoresby served as an apprentice in his

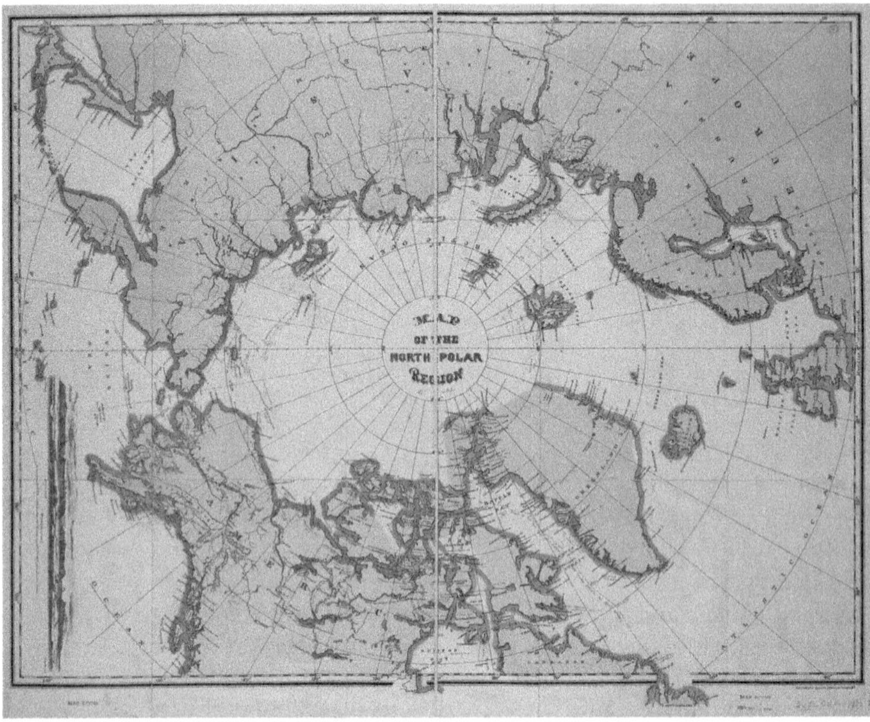

FIGURE 7 Map of the North Polar Regions, 1882.
From The Lionel Pincus and Princess Firyal Map Division. The New York Public Library

father's employ and at age sixteen he was promoted to first mate. Scoresby had enjoyed much success in 1811, rewarding the faith placed in the twenty-one year old novice captain by the owners of the *Resolution*.[1]

[1] The scientific name for the bowhead whale is *Balaena mysticetus*. Although the word bowhead was not used by the British, it is employed throughout as it has come to be the most common name for *Balaena mysticetus*. First used by American whalemen in the nineteenth century, the name bowhead is a reference to the animal's bow-shaped skull. J. Jerome Montague, "Introduction," in *The Bowhead Whale*, ed. John J. Burns, J. Jerome Montague, and Cleveland J. Cowles (Lawrence, KS: The Society for Marine Mammalogy, 1993), 1–21. For a brief biographical sketch of William Scoresby, see C. Ian Jackson, ed., *The Arctic Whaling Journals of William Scoresby the Younger: Volume I, The Voyages of 1811, 1812 and 1813* (London: The Hakluyt Society, 2003), xxi–xxxiii. Scoresby's career as whaling captain, scientist, and cleric is also addressed in Michael Bravo, "Geographies of Exploration and Improvement: William Scoresby and Arctic Whaling, 1782–1882," *Journal of Historical Geography* 32 (2006): 512–38; Anita McConnell, "The Scientific Life of William Scoresby, Jnr., with a Catalogue of his Instruments and Apparatus in the Whitby Museum," *Annals of Science* 43 (1986): 257–86; and Constance Martin, "William Scoresby, Jr. (1789–1857) and the Open Polar Sea – Myth and Reality," *Arctic* 41, no. 1 (March 1988): 39–47. The most recent full biography of Scoresby is by Tom Stamp and Cordelia Stamp, *William Scoresby, Arctic Explorer* (Whitby, England: Caedmon of Whitby Press, 1975). For Scoresby's 1812 preparations, see Jackson, *The Arctic Whaling Journals of William Scoresby*, 63–6.

FIGURE 8 Rev.d William Scoresby, 1824.
Print Collection portrait file. /S/ William Scoresby (1789–1857). 1824.
Print Collection, Miriam and Ira D. Wallach Division of Art, Prints and Photographs, The New York Public Library, Astor, Lenox, and Tilden Foundations

The timing of the *Resolution's* March 27th departure from Whitby for the 1812 Greenland whaling season was dictated by a confluence of factors relating to meteorology, oceanography, glaciology, early nineteenth-century maritime technology, and the patterns of bowhead migration. As Scoresby explained, the conditions prior to the middle of April in the areas north of 75° latitude where bowheads might be found were simply too dangerous. The risk of encountering the prevalent drift ice in the dark prior to the spring advent of twenty-four hour daylight far outweighed any advantage an early start to the whaling season might provide.

FIGURE 9 Vessel in ice with icebergs behind at left.
Courtesy of the New Bedford Whaling Museum

The severity of the frost and storms before the middle of April in combination with the ice and darkness "probably produce as high a degree of horror in the mind of the navigator . . . as any combination of circumstances which the imagination can present." If Scoresby's fellow captains were wary of the conditions they might encounter if they reached the whaling grounds too early, they were also concerned that their competitors might beat them to the whales. A captain with a head start over his colleagues might in a single week alone amidst a herd of bowheads guarantee a profitable season before other masters even spied a single whale.[2]

The gamesmanship of captains in timing their departure from Whitby was a mere prelude to the constant interactions between masters that marked any season in a whalefishery. Whaling was, fundamentally, a group endeavor. Intensely competitive, yet, at the same time, whalemen were remarkably collegial and, on occasion, extremely generous. Captains were well aware that their success was dependent upon watching the movements of other ships through the ice and in sharing information about conditions and bowhead locations with their fellow masters. Scoresby, like all whaling captains, recognized that the master, whose miscalculation of ice conditions could be taken advantage of to gain a whale, might later prove savior should the *Resolution* find itself trapped in the ice. As competitors and allies in a shared battle in a harsh environment where the flick of a bowhead's tail or the tardy cutting of a line

[2] William Scoresby, *An Account of the Artic Regions: With a History and Description of the Northern Whale-Fishery* (Edinburgh: Archibald Constable and Co., 1820), 2:205–11, 207.

attached to a sounding whale could mean death, how did British whalers resolve ownership disputes over their valuable quarry?[3]

Scoresby was very clear, in his magisterial 1820 *An Account of the Arctic Regions with a Description of the Northern Whale-Fishery*, that whalemen had, in the absence of legislation, created their own "equitable system of regulations" that curbed the greed of individual captains and worked to the "mutual benefit" of all. He explained that the Greenland practice of fast-fish, loose-fish discouraged litigation and prevented whales from escaping capture. Scoresby indicated that when a whale was struck by a harpoon it became the property of the ship that dealt the blow. Possession was retained so long as the whaler maintained that connection through the harpoon and lines to the animal. It did not matter that the line so tenuously rested upon the whale that its escape could be easily achieved. If, at the moment after a second whaler inflicted a fatal wound, the line of the original hunter should lose its connection to the whale the first striker would still be entitled to possession. Upon harpooning a whale, a whaleboat, and the ship from which it was sent hoisted a flag, or jack, to signal friends for assistance and to warn off competitors. A whaler flying the jack was presumed to remain fast to a whale unless the second ship could present convincing evidence to the contrary. The competitor's claim was always subject to the compelling rejoinder that a displaced line may well have maintained contact with the whale beneath the water's surface and out of view.[4]

If the fast-fish provisions favored the original striker, the loose-fish component worked to the advantage of an intervening whaler. Once a whale was loose, regardless of the circumstances, it was free for the taking. The harshness of this rule was illustrated by Scoresby's hypothetical example of a ship towing a dead whale awaiting the passing of a storm to commence cutting in. If the line should break, a second whaler would be free to claim the animal. Scoresby acknowledged that this custom may appear inequitable. He defended the Greenland practice, however, with the observation that it prevented whales from going uncaptured and diminished the likelihood of litigation. If the law permitted a ship to maintain possession of a whale to which it was no longer connected, the contentious issue would arise as to what constituted a sufficient degree of proximity and pursuit for the original striker to retain possession. A second whaler would, under this framework, hesitate to pursue a loose whale out of fear that its effort would go unrewarded or result in expensive litigation should it capture the animal. The end result, Scoresby explained from his perspective as a captain, would be unfortunate: the failure of the Greenland fishery to yield its maximum economic return.[5]

Yet, the universal respect for these principles established by custom in the Greenland fishery seems to evaporate as Scoresby proceeded to explain their

3 Scoresby, *An Account of the Arctic Regions*, 2:340–68, 446–9; and Jackson, *The Arctic Whaling Journals of William Scoresby*, 71.

4 Scoresby, *An Account of the Arctic Regions*, 2:312, 318–28, 522–3.

5 Scoresby, *An Account of the Arctic Regions*, 2:318–28.

actual application at sea. What, at first glance, appears simple and certain becomes a good deal messier in practice. After extolling the practical, legal virtues of fast-fish, loose-fish, Scoresby suggested that rules of whaling would be improved by observance of the precept from *Matthew* 7:12 that one do unto others. While this appears, at first, to be the pious platitude of a man who in a few years would seek ordination, it was a compelling argument for a new standard. Scoresby asserted that to take a whale bearing the harpoon, but not the line, of another ship constituted robbery and "that nothing could justify such an act, but the certainty, that the original strikers could have little or no chance of ever recovering the possession of the fish itself." Particular note was taken of a captain who sends his boats to assist in the taking of a whale fast to a competitor. Should that whale become loose, Scoresby asserted that "according to the principles of right and strict honor, I conceive it is the property of the first striker, and as such, ought to be given up." Having provided several examples of the unfairness of fast-fish, loose-fish, Scoresby – despite his obvious preference for a rule that allowed a harpoon to mark a whale for future capture – lamented that the economic self-interest of whalemen clouded the "integrity of character" needed to permit adoption of a new rule. An explanation for Scoresby's obviously conflicted feelings on this topic and the degree to which the rules of the Greenland fishery were not nearly as settled as the captain suggested, can be found in the logbook Scoresby kept during the 1812 voyage of the *Resolution*.[6]

On July 21, 1812 at about 78° north and 4° east in the prime bowhead grounds near Spitzbergen, the crew of the *Resolution* was determined to profitably spend their final few days in the fishery before returning home. The *Resolution*, having followed the *John* through some ice, encountered several bowheads. One of the boats belonging to the *Resolution* attempted to strike a whale, but its harpoons failed to stay fast. Disappointed, Scoresby ordered his boats to assist the *John* in capturing a fast whale that was running their competitor's line toward the *Resolution*. Scoresby's decision to assist the *John* would not have been viewed by his crew as an unusual order. Captains often directed that such assistance be rendered. While such aid was not rewarded with a share of the captured whale, the help was recognized as part of a code of behavior that urged recipients of assistance to reciprocate the favor to another vessel. Had the *John's* line remained fast, any harpoons from the *Resolution* that found their mark would have been deemed "friendly" and returned to the latter with a show of appreciation when the target of the chase was captured. As frequently happened, however, the *John's* harpoon drew loose. The whale that had for the previous hour been pursued by the *Resolution* on behalf of the *John* was now, according to a strict interpretation of fast-fish, loose-fish, free for the taking. The *Resolution* managed thereafter to affix a harpoon and take the whale.[7]

[6] Scoresby, *An Account of the Arctic Regions*, 2:318–28, 326–7.
[7] Jackson, *The Arctic Whaling Journals of William Scoresby*, 119–20.

As a twenty-two year old captain in only his second year in command, Scoresby faced a serious challenge. His crew was adamant that pursuant to the custom of the fishery the whale was rightfully the sole property of the *Resolution*. Scoresby concurred that they had a legal right to the whale, but argued that "by the laws of honor" the claim of the *John* should prevail. The *Resolution*, after all, had rendered aid with the intent that the *John* should take the prize. Scoresby's men would never have been in a position to take the whale without the actions of the *John* in originally striking and wounding the creature. The crew of the *Resolution* countered with the rejoinder that the *John* had fallen off the pace of the whale once it swam free of its harpoon and would never have successfully closed the gap. Scoresby's crew in arguing for a strict application of fast-fish, loose-fish introduced an essential element of iron holds the whale: the likelihood that the *John* would, without the assistance of the *Resolution*, have captured its quarry.[8]

Scoresby's dilemma was complicated by the fact that his father was the captain of the *John*. Revered as one of England's leading and most innovative whalemen, William Scoresby, Sr. was an imposing figure. In addition to the normal familial bonds, the younger Scoresby almost certainly owed his command to his father's influence. As the crew of the *Resolution* was well aware, the father had passed the ship's command to his son when he assumed the captaincy of the *John* the previous year. The discomfort felt by the son is evident in the pages of the *Resolution's* logbook. "In this dilema [sic] placed between the duty as a Master and duty as a Son which had opposite actions I was in an unhappy strait." A similar conflict with any other vessel would have permitted Scoresby to do what he considered honorable and surrender the whale free of the accusation that he was acting in the thrall of his father's influence. In a move that might have struck his crew as desperate and undermined his authority, Scoresby argued that his father was "wroth." That the captain of the *John* was extremely angry meant little to the crew of the *Resolution* convinced that the whale was rightfully theirs. In an attempt to resolve the dispute to the satisfaction of all, Scoresby boarded the *John* to speak with his father.[9]

Scoresby's initial suggestion was that they simply split the whale. The elder Scoresby was not immediately disposed to follow this course. After much

[8] Jackson, *The Arctic Whaling Journals of William Scoresby*, 119–20.

[9] Jackson, *The Arctic Whaling Journals of William Scoresby*, 119–20. For Scoresby's own assessment of his father's ability to intimidate others through his physical strength and forceful personality, see William Scoresby, *Memorials of the Sea, My Father: Being Records of the Adventurous Life of the Late William Scoresby, Esq. of Whitby* (London: Longman, Brown, Green, and Longmans, 1851), 21–3. The younger Scoresby did not make reference to his unpleasant 1812 encounter with his father when he penned a biography of the latter nearly forty years later.

discussion, the *John's* master agreed, or so it appeared. While the logbook account is a bit cryptic, the elder Scoresby apparently reverted suddenly to his earlier state of rage. He blamed his son for the entire situation and threatened to enforce his legal right to the disputed whale once he returned to England. Disconsolate, Scoresby returned to the *Resolution* where the crew began removing the whale's blubber for storage. Likely seeking some sort of reconciliation with his father, Scoresby thereafter had his ship draw up close to the *John*. The logbook entry for the day ends with a statement that described the ship's position and captured the mood of its author. "The Ice seems to have quite enclosed us."[10]

This incident, like most disputes over captured whales, did not end in litigation. Was the elder Scoresby's threat of legal action, therefore, merely empty invective shouted in frustration at a lost whale or was it made in the hope that the courts might provide some relief? A quick review of the reported British cases might suggest that Scoresby's father knew full well that any court would award the whale as a loose-fish to the *Resolution*. As early as 1786, a dispute between two British whalers was resolved by application of fast-fish, loose-fish. *The Times* of London reported that Justice Buller of the Court of King's Bench instructed the jury that according to the law of the Greenland whalefishery any fast whale that "gets loose by any means, though she has a harpoon and lines about her," may be taken by another ship. At the York Lent Assizes in 1788 a ruling was issued in *Littledale v. Scaith* that provided a clear statement of fast-fish, loose-fish that came to be cited by most judges and legal commentators throughout the nineteenth century as the universally accepted custom of the Greenland fishery. Yet, the elder Scoresby's threat of litigation was not made without some basis in the law and the practices of the Greenland fishery. Despite frequent speculation in the nineteenth century that the custom as set forth in *Littledale* had been followed since the dawn of Greenland whaling and the tendency of many courts and commentators to view fast-fish, loose-fish as rigid and unchanging, it was, in fact, a fairly recent development in the late eighteenth century and – as Scoresby's experience suggests – was never uniformly applied at sea.[11]

[10] Jackson, *The Arctic Whaling Journals of William Scoresby*, 119–20.

[11] The 1786 matter before Justice Buller was not preserved in one of the volumes of reported cases from which lawyers and judges in common law jurisdictions find precedents. *Times* (London), December 29, 1786. For a brief account of Justice Francis Buller's career, see William C. Townsend, *The Lives of Twelve Eminent Judges of the Last and of the Present Century* (London: Longman, Brown, Green, and Longmans, 1846), 2 volumes, I: 1–32. *Littledale v. Scaith*, 1 Taunt. 241, 127 Eng. Rep. 825 (1788). *Littledale* is appended as a note in the reported Galapagos Island whaling dispute *Fennings v. Lord Grenville*, 1 Taunt. 241, 127 Eng. Rep. 825 (1808). For speculation as to the antiquity of Greenland custom, see, for example, the argument of counsel in *Fennings*, 1 Taunt. 243–4.

II

Spitzbergen or Svalbard – as it is known at present – is an archipelago approximately half way between Norway and the North Pole. Although the Vikings possibly knew of its existence, the Dutch explorer William Barents discovered in 1596 an island he dubbed Spitzbergen. In 1607, the English explorer Henry Hudson sailing under a Dutch flag further investigated the island's coastline. What Barents and Hudson had reached was the largest and nearly westernmost island of the archipelago. About 280 miles long and anywhere from 25 to 140 miles in width, the island, also sometimes called West Spitzbergen, is situated along an extension of the Gulf Stream that keeps the icy waters of its western coast open to navigation approximately six months each year. Although the investigation of Spitzbergen did nothing to advance his goal of finding the hoped for Northeast Passage to the Pacific, Hudson recognized the commercial possibilities of the region. Spitzbergen's many harbors abounded with bone-laden, and oil-rich bowhead whales. European commercial whaling heretofore was mostly shore-based and generally limited to Basque hunters primarily off the beaches of southwest France and, secondarily, North America's northeast coast. The abundant bowheads in the bays off the west coast of Spitzbergen were ideal for the prevailing whaling technique that entailed spotters on beach towers alerting crews of generally six men to put to sea at a moment's notice in pursuit of likely prey. Similar crews in other nearby small boats were then summoned to assist in the harpooning and lancing of the targeted whale.[12]

[12] For Barents and Hudson and their discovery of Spitzbergen, see Louwrens Hacquebord, Frits Steenhuisen, and Huib Waterbolk, "English and Dutch Whaling Stations in Spitsbergen (Svalbard) before 1660," *International Journal of Maritime History* XV, no. 2 (December 2003): 117–34; John J. Teal, "Europe's Northernmost Frontier," *Foreign Affairs* 29, issue 2 (January 1951): 263–75; Robert C. Allen and Ian Keay, "Bowhead Whales and the Eastern Arctic, 1611–1911: Population Reconstruction with Historical Whaling Records," *Environment and History* 12 (2006): 89–113; Robert C. Allen and Ian Keay, "Saving the Whales: Lessons from the Extinction of the Eastern Arctic Bowhead," *Journal of Economic History* 64, issue 2 (June 2004): 400–32; Robert C. Allen and Ian Keay, "The First Great Whale Extinction: The End of the Bowhead Whale in the Eastern Arctic," *Explorations in Economic History* 38 (2001): 448–77; and Lance E. Davis, Robert E. Gallman, and Karin Gleiter, *In Pursuit of Leviathan: Technology, Institutions, Productivity, and Profits in American Whaling, 1816–1906* (Chicago: University of Chicago Press, 1997), 31–4. For Basque whaling, see Selma Huxley Barkham, "The Basque Whaling Establishments in Labrador 1536–1632 – A Summary," *Arctic* 37, no. 4 (December 1984), 515–9; Aldemaro Romero and Shelly Kannada, "Comment on 'Genetic Analysis of 16th-Century Whale Bones Prompts Revision of the Impact of Basque Whaling on Right and Bowhead Whales in the Western North Atlantic,'" *Canadian Journal of Zoology* 84, no. 7 (2006), 1059–65; and Eric Jay Dolin, *Leviathan: The History of Whaling in America* (New York: W. W. Norton & Co., 2007), 21–4. For the basic geography of the archipelago, see CIA World Factbook, s.v. "Svalbard," accessed June 28, 2015, www.cia.gov/library/publications/the-world-factbook/geos/sv.html. Spitzbergen has retained a dual meaning. It can refer to both the largest island and the entire archipelago. As English whalemen in seventeenth and eighteenth century generally spelled the island "Spitzbergen," rather than the more modern "Spitsbergen," the older spelling has been employed throughout.

Dutch, British, German, and Basque operations were quickly established, setting off in the first years of the fishery contentious battles – occasionally with the use of arms – for the best harbors. Shore-based whaling with its use of buildings to boil blubber, store oil, and house workers precluded the creative, spontaneous decision making that so marked a successful pelagic captain. Anchored to a particular location for an entire season, shore-based whalemen displayed a territoriality absent once ships ventured out to sea for extended periods of time. As whaling historian Gordon Jackson has remarked, the English effort in the first years of Spitzbergen whaling was directed more at excluding the Dutch than actually killing whales. Violence and the threat of further disruptions in the business of harvesting oil led the participating nations prior to the 1619 season to an agreement assigning Spitzbergen harbors by nationality. While the 1619 assignment of harbors was initially viewed as a victory for the British, the northern coast of Spitzbergen to which the Dutch were relegated proved particularly advantageous. The annual migration of bowheads from the north allowed the Dutch to thin the ranks of whales available in the English harbors in the south. Dutch whalemen also used their northern base of operations to venture out to the edge of the ice in the vicinity of 79° north. This allowed the Dutch to gain their first experience of pelagic whaling in an area that would prove to be the Greenland fishery's most productive hunting grounds. By 1650, the Dutch began a century as the pre-eminent European whaling nation.[13]

[13] Hacquebord, Steenhuisen, and Waterbolk, "English and Dutch Whaling Stations in Spitsbergen," 129–33; Gordon Jackson, *The British Whaling Trade* (Hamden, CT: Archon Books, 1978), 11–26; Allen and Keay, "Bowhead Whales and the Eastern Arctic, 1611–1911," 90–5; Allen and Keay, "Saving the Whales," 400–1; Allen and Keay, "The First Great Whale Extinction," 448–51; and Davis, Gallman, and Gleiter, *In Pursuit of Leviathan*, 31–4. The Dutch ultimately took control of the Greenland fishery for a number of reasons. For the advantages the Dutch enjoyed in labor productivity in whaling, see Christiaan van Bochove and Jan Luiten van Zanden, "Two Engines of Early Modern Growth? Herring Fisheries and Whaling during the Dutch Golden Age (1600–1800)," in *Ricchezza del Mare secc XIII–XVIII*, ed. S. Cavaciocchi (Grassina, Italy: Le Monnier, 2006), 557–74. While dealing with issues of productivity for other areas of Dutch maritime activity, the following unpublished draft article is suggestive for the whaling industry: Milja van Tielhof and Jan Luiten van Zanden, "Productivity Changes in Shipping in the Dutch Republic: The Evidence from Freight Rates, 1550–1800," accessed January 17, 2008, www.iisg.nl/research/two-engines.pdf and www.iisg.nl/research/productivity-shipping-app.pdf. For the ability of Dutch merchants to shape government policy in ways favorable to their interests, see Jelle C. Riemersma, "Oceanic Expansion: Government Influence on Company Organization in Holland and England (1550–1650)," *The Journal of Economic History* 10 (1950), 31–9; and J. Braat, "Dutch Activities in the North and the Arctic during the Sixteenth and Seventeenth Centuries," *Arctic* 37, no. 4 (December 1984), 473–80. For Jackson's view of English efforts to defeat Dutch whaling interests, see Gordon Jackson, "Government Bounties and the Establishment of the Scottish Whaling Trade, 1750–1800," in *Scottish Themes: Essays in Honour of Professor S. G. E. Lythe*, ed. John Butt and J. T. Ward (Hamden, CT: Archon Books, 1976), 47. The Dutch asserted that the waters near Spitzbergen constituted *mare liberum* and thus the bowheads in the archipelago's bays were not the property of any one

The Dutch, unlike the British and later the Americans, provided some statutory guidance concerning disputed whales. The States-General of Holland and West Friesland enacted in 1695 a comprehensive set of laws governing its whaling industry. Each captain and chief officer was required to take an oath prior to departure swearing adherence to the regulations. Revising an unsuccessful code promulgated in 1677, the new law provided among its twelve articles:

9. Any one having killed a whale in the ice, but who cannot conveniently take it on board, shall be considered as the owner thereof, so long as any of his crew remains along with it; but whenever it is deserted, though made fast to a piece of ice, it becomes the property of the first who can get possession of it.

10. But if a fish be made fast to the shore, or moored near the shore by means of a grapnel or anchor, with a buoy, a flag, or other mark attached to it, signifying that it is not deserted, – the person who left it there, shall still be considered the sole proprietor, though no person may be with it.

The Dutch law reflects the mid to late seventeenth-century transition from shore-based to pelagic whaling. Both provisions govern situations involving a whale that has been killed, but for some reason cannot be immediately taken back to the ship or shore to be rendered. The burden was on the successful whaler to demonstrate to other hunters his control over the carcass. Unless the whale was anchored close to shore and marked with a flag or some other sign of possession, a crew member must remain with the body. Marking a whale that was fast to the ice with a buoy or flag was not sufficient. The requirement that a whale attached to ice have a human companion was perhaps an acknowledgment that drifting ice may prevent the boat that inflicted the mortal blow from finding and reclaiming the animal. The logic was likely that it was better that the efforts of one whaler go unrewarded than that a valuable commodity go unclaimed by a second whaler who, seeing a flagged whale at sea, allowed it to drift off. The presence of a crew member was, in effect, insurance that a boat would return for both man and whale. The anchoring of a dead whale near the beach – a common practice in shore-based whaling – required no such assurance of the captor's return. In the shallow water near to whaling stations there was no danger that an anchored whale would escape utilization by whalemen who retained a strong sense of territorial possession of particular bays and harbors.[14]

nation. The English argued – somewhat tortuously – that they had established sovereignty over Spitzbergen and its bowheads. For the legal disputes over Svalbard which, since 1920, has been under Norwegian control, see Torbjørn Pedersen, "The Svalbard Continental Shelf Controversy: Legal Disputes and Political Rivalries, *Ocean Development & International Law* 37 (2006), 339–58.

[14] Scoresby, *An Account of the Artic Regions*, 2:316–7.

III

The Dutch provisions did not cover the familiar situation at sea where two or more ships were actively pursuing a live whale. The resolution of such disputes in the late seventeenth century appears to have been framed in the terms of the ancient debate between Roman lawyers over ownership of wild animals or what they termed *ferae naturae*. Roman authorities such as Trebatius and Gaius agreed that the mere pursuit of a wild animal did not confer ownership. The pursuer must possess the creature. Scholars differed, however, as to what constituted possession. Gaius argued that nothing less than the actual physical possession of the beast was required to gain title. Trebatius, on the other hand, asserted that possession vests prior to capture if the first pursuer inflicts a severe wound and maintains a pursuit that leaves the animal little chance of escape. In such a situation the creature has been deprived of its freedom and brought within the power of the hunter. James Dalrymple, the first Viscount of Stair, illustrated his 1681 discussion of *ferae naturae* with the observation that the notion of possession advanced by Trebatius prevailed in the Greenland whalefishery. Stair explained that the Greenland whalemen who "woundeth a whale so that she cannot keep the sea for the smart of her wound, and so must needs come to land, is proprietar, and not he that lays first hand on her at land."[15]

Stair's discussion of possession in the Greenland fishery was not limited to whales that drift ashore. He indicated that a similar custom was applied to whales that were pursued out at sea. A whalemen in sole pursuit of a whale with "a probability to reach his prey" was awarded the prize even if another vessel ultimately captured the animal. Stair illustrated Greenland custom by reference to a 1677 case decided by the Court of Session in Scotland. During the Anglo-Dutch Wars, an English frigate and an allied French vessel captured a Dutch privateer in possession of three prizes. When two additional Dutch ships attempted to rescue their countrymen, one of the prize vessels, the *Tortoise*, attempted to make its escape. Having chased the Dutch interlopers away, the English and French frigates turned their attention to recovering the *Tortoise*. A Scottish privateer intervened, however, and captured the *Tortoise*. The resulting litigation over the prize was decided by the Court of Session in favor of the Scottish privateer. Stair explained that the Scottish high court – applying Trebatius' rationale concerning possession of a wild animal – determined that despite its damaged condition the *Tortoise* would have made good its escape if not for the actions of the Scottish privateer. Had the evidence shown that the

[15] Stair, James Dalrymple, *Viscount of, The Institutions of the Law of Scotland, Deduced from its Originals, and Collated with the Civil, and Feudal-Laws, and with Customs of Neighboring Nations* (Edinburgh: G. Hamilton and J. Balfour, 1759), Book II, Title I, Section XXXIII, 179–80. For a wide ranging discussion of the law and wild animals, see Steven M. Wise, "The Legal Thinghood of Nonhuman Animals," *Boston College Environmental Affairs Law Review* 23, issue 3 (Spring 1996): 471–546.

English frigate was in position, absent interference, to capture the prize, the result would have been different.[16]

The custom of the Greenland whalefishery, as explained by Stair, was not entirely clear. How certain, for example, must the capture of a whale by the first striker be before ownership will be awarded? Stair indicated at one point in his discussion that a mere probability was sufficient, yet he also related that the Scottish privateer prevailed because the English frigate failed to prove – an apparently higher standard – that absent interference it would have captured the *Tortoise*. Issues of the applicable standard of proof aside, Stair also suggested that a whale need not be seriously injured by the pursuing first striker in order to gain possession prior to the actual capture of the animal. The pursuit of a whale, coupled with the probability of success, appeared – in some circumstances – sufficient to secure ownership.

Confirmation that the fast-fish, loose-fish requirement that a physical connection be maintained between whale and boat or crew was not universally observed since the origins of the Greenland fishery was provided by a 1778–1779 geography text and a 1792 ruling by the Scottish Court of Session. Both sources further revealed that the practice of iron holds the whale was also observed in late eighteenth-century Greenland and was not, as Robert Ellickson speculates, an adaptation to hunting swifter and more combative sperm whales. Charles Theodore Middleton in his 1778–1779 work explained that in the Greenland fishery whales occasionally break free from an attached line. "When this happens, however, if he is afterwards taken by the crew of another ship," Middleton revealed, "he is returned to those who first wounded him, as that is known by the harpoon, which is always distinguished by a peculiar mark." In *Addison v. Row*, the *Priscilla*, an English ship in the Davis Strait, struck a whale and raised a flag to signal that it was fast. A nearby Scottish vessel, the *Caledonia*, lowered four boats to assist the *Priscilla* in capturing its prey. When the seriously injured whale emerged from its dive, the *Caledonia* promptly killed the animal that still bore the *Priscilla's* harpoon. Rather than return the whale as the crew of the *Priscilla* expected, the *Caledonia* kept the animal, arguing that it had broken free from its original attacker prior to being dealt the fatal blow. The *Caledonia's* claim was a clear invocation of the fast-fish, loose-fish rule that whales not in contact with its pursuers were free for the taking. The Court of Session did not concur, advancing, instead, a rule very much akin to iron holds the whale. Lord President Campbell proclaimed that pursuant to the "general rule" the whale "belongs to the first occupant, being naturally *res nullius*. But if I once seize upon the animal, and it breaks away from me, and I still continue in pursuit, I do not thereby lose my right as first occupant, so

[16] Stair, *The Institutions of the Law of Scotland*, Book II, Title I, Section XXXIII, 179–80. *The King's Frigate against a Scots Caper*, Advocates' MS. No. 535, folio 273 in M. P. Brown, ed., *Supplement to the Dictionary of the Decisions of the Court of Session* (Edinburgh: W. & C. Tait, 1826), 3:125.

long as there are hopes of recovering it. There is no custom proved which can derogate from this general principle. . . . The boats of the Priscilla would have taken the whale if the *Caledonia* had never interfered."[17]

The *Priscilla's* victory was short-lived. The House of Lords, to which Scottish cases since the Union of 1707 had been subject to appeal, viewed the law in starkly different terms. Lord Thurlow stated, "It is a settled point, that a whale being struck, and afterwards getting loose, is the property of the next striker who continues fast till she is killed." Applying fast-fish, loose-fish to the evidence adduced before the Scottish court, the House of Lords ruled that when the *Caledonia* struck the whale it was loose and free for the taking. The case of *Addison v. Row* thereafter joined *Littledale v. Scaith* and the 1786 decision of Justice Buller in establishing in the minds of British jurists and lawyers fast-fish, loose-fish as the universally accepted custom of the Greenland fishery. Subsequent decisions such as the 1805 ruling in *Gale v. Wilkinson* which was cited by Scoresby and Melville solidified this perception. The question thus arises as to whether Lord Campbell was mistaken in finding that fast-fish, loose-fish had not supplanted what he viewed as the general rule of honoring the rights of a first striker that remained in pursuit with reasonable prospects for success? Middleton clearly established that Lord Campbell was not mistaken in recognizing that at least some Greenland whalemen in the late eighteenth century did not believe that a physical connection to a whale was required to constitute possession. It is impossible, of course, to know the degree to which particular norms were practiced in Greenland. The power of the House of Lords to overturn the Scottish decision established the ascendancy of fast-fish, loose-fish and quickly erased the memory of iron holds the whale as a custom once observed in the Greenland fishery.[18]

[17] Charles Theodore Middleton, *A New and Complete System of Geography: Containing a Full, Accurate, Authentic and Interesting Account and Description of Europe, Asia, Africa, and America* (London: J. Cooke, 1778–1779), 2:22. *Addison v. Row*, in *Reports of Cases Decided in the House of Lords upon Appeal from Scotland, from 1753 to 1813*, ed. Thomas S. Paton (Edinburgh: T. & T. Clark, 1853), 3:334–40, 338. Although the Davis Strait separates Baffin Island from the west coast of Greenland, British whalemen and government officials considered it part of the Greenland whalefishery. Government bounties for Greenland whaling included the waters between Davis Strait and Spitzbergen. Eighteenth- and nineteenth-century keepers of British whaling statistics frequently combined the fertile grounds of Spitzbergen and Davis Strait into a single Greenland fishery. Other compilations separate entries for Davis Strait from a Greenland fishery defined as the area east of Greenland. For a reliable general introduction to British whaling, see Jackson, *The British Whaling Trade*. For Ellickson's speculation as to the importance of the type of whale hunted and the applicable custom, see Robert C. Ellickson, "A Hypothesis of Wealth-Maximizing Norms: Evidence from the Whaling Industry," *Journal of Law, Economics and Organization* 5 (Spring 1989): 83–97, 89–92; and Robert C. Ellickson, *Order without Law: How Neighbors Settle Disputes* (Cambridge: Harvard University Press, 1991), 197–201.

[18] *Addison v. Row*, in *Reports of Cases Decided in the House of Lords*, 3:339. For a discussion of the House of Lords hearing appeals from the Court of Session after 1707, see David M. Walker,

The conflicting decisions rendered by the Scottish Court of Session and the House of Lords in *Addison* are indicative of some of the important differences between Scottish and English law in the late eighteenth century. While a contentious debate between legal historians as to the degree to which Scottish law is a mix of the civilian tradition and English common law persists, it is beyond peradventure that Roman law shaped the way courts north of the River Tweed viewed the creation of law. Roman law was never accepted, in full, as constituting the common law of Scotland. It did, however, form an intellectual foundation and a structure for Stair, George Mackenzie, John Erskine, and the other so-called Institutional writers of the seventeenth and eighteenth centuries in their attempts to set forth the nature of Scottish law. The authority enjoyed by Roman law was, in large part, a product of the type of legal education open to aspiring Scottish lawyers prior to 1750. Educated mostly in France and later Holland, generations of Scottish law students returned home with a deep appreciation of Roman law and the civil law tradition. Erskine asserted in the posthumous 1805 edition of *An Institute of the Law of Scotland* that of "all the systems of human law which now exist, the Roman so well deserves the first place, on account of the equity of its precepts, and the justness of its reasonings." Erskine illustrated the extent to which his view was shared by his countrymen with the observation that the pre-1707 Estates of Parliament frequently justified legislation by declaring it conformable to Roman law. Roman law also provided guidance for situations not clearly governed by statutes or custom. The paucity of Scottish legislation, Erskine explained, further magnified the influence of Roman law.[19]

"Some Characteristics of Scots Law," *The Modern Law Review* 18, no. 4 (July, 1955): 321–37. For *Gale v. Wilkinson*, see *Times* (London), December 24, 1805; Herman Melville, *Moby-Dick or, The Whale* (New York: Penguin Books, 2003), 433–4; and Scoresby, *An Account of the Arctic Regions*, 2:323–4, 518–21. Scoresby supplied a report of the case of *Gale v. Wilkinson* as an appendix to his book. In *Gale*, a boat from the *Neptune* struck a whale at the edge of an ice field in the Spitzbergen fishery. The harpoon line being soon fully extended, the crew abandoned the boat for the safety of the ice and allowed the whale to carry it under the surface. When the whale reappeared with the boat still attached, a second ship, the *Experiment*, arrived and, with the assistance of the *Neptune's* boats, captured the whale. The *Neptune's* expectation that the whale would be returned was dashed when the captain of the *Experiment* refused to even return the attached harpoons, lines, and boat. In the ensuing litigation in England, Lord Ellenborough ruled that the *Neptune* had lost possession of the whale when its crew abandoned the attached whaleboat. It did not matter what was attached to the whale. The crucial issue was the contact between the whale and the crew of the *Neptune*. Lord Ellenborough did direct that the boat be returned to the *Neptune's* owner. In the most famous part of the decision, the line and the harpoon were retained by the *Experiment* pursuant to the theory that the whale had obtained a sort of property right to this equipment that was transferred to the *Experiment* upon capture of the animal. The law, while often lacking a sense of humor, is frequently amusing.

19 John Erskine, *An Institute of the Law of Scotland, in Four Books: In the Order of Sir George Mackenzie's Institutions of that Law* (Edinburgh: Bell & Bradfute, 1805), 1:1–22, 9. For an introduction to some of the differences between Scottish and English law in this period, see W. D. H. Sellar, "Scots Law: Mixed from the Very Beginning? A Tale of Two Receptions,"

The importance of the civil law tradition in the development of Scottish law was not limited to substantive Roman law. Scottish legal scholars also imbibed on the continent a particular way of thinking about law and the resolution of legal problems foreign to those schooled in English common law. The civil law tradition – of which Roman law is the primary component – encourages its practitioners to grasp the large, foundational principles behind what might otherwise appear to be a jumble of disconnected statutes and customs. The law, understood in this way, is about deducing answers that advance, in Erskine's words, "an equal distribution of justice, on which the happiness of every society depends." As the legal system deemed most steeped in justice, Roman law naturally provided an invaluable guide to moving from broad principle to a particular application.[20]

English common law, on the other hand, was based in this period on the fiction that it gained much of its authority from immemorial usage. Provisions that make up the common law are, Blackstone famously remarked, customs the origins of which must date to a "time whereof the memory of man runneth not to the contrary." Evidence as to the customs that make up the common law of England is found in the reports of previous decisions rendered by generations of learned judges who, in their study of precedent, are "living oracles" of English law. An English judge is duty bound to follow precedent in deciding a case even if the logic of the established custom is not immediately apparent. Blackstone indicated that only a precedent that is flatly absurd or unjust need not be followed. The truth, of course, is that the theoretical rigidity of custom has always encouraged a rather elastic application of the concept. English jurists in the late eighteenth century – like judges in all common law jurisdictions – were adept at distinguishing the facts in the case at bar from those in a troubling precedent to avoid an unfortunate result. Judges were also willing to confer common law status where a practice was of decidedly recent origins.[21]

The problem of strictly applying the English common law notion of custom in a whaling dispute is well illustrated by the 1808 Common Pleas decision in

Edinburgh Law Review 4 (2000): 3–18; J. T. Cameron, "Custom as a Source of Law in Scotland," *Modern Law Review* 27 (1964): 306–21; John W. Cairns, "Attitudes to Codification and the Scottish Science of Legislation, 1600–1830," *Tulane European and Civil Law Forum* 22 (2007): 1–78; and Walker, "Some Characteristics of Scots Law," 321–37. For the education of Scots lawyers, see Cairns, "Attitudes to Codification and the Scottish Science of Legislation," 4–5.

[20] Erskine, *An Institute of the Law of Scotland*, 1:2.

[21] E. K. Braybrooke, "Custom as a Source of English Law," *Michigan Law Review* 50 (1951): 71. For Blackstone's discussion of custom, precedents, and judges, see William Blackstone, *Commentaries on the Laws of England* (London: T. Cadell, 1793), 63–85, 67, 69. Cambridge professor Edward Christian countered in his comments appended to the 1793 edition of Blackstone's *Commentaries* that even flatly absurd or unjust provisions must be followed as properly constituted English law so long as they were in general accord with the long standing principles of English law. The remedy in such situations must be had in Parliament. Blackstone, *Commentaries on the Laws of England*, 70.

Fennings v. Lord Grenville. In 1805, two English ships, the *William Fennings* and the *Caerwent*, were hunting for whales in the vicinity of the Galapagos Islands. The *William Fennings*, while killing a whale, struck a second whale. A small buoy, or drogue, was attached to the second whale to slow its progress and mark its position for later capture. The *Caerwent* subsequently took up the chase and killed the whale. All agreed that pursuant to fast-fish, loose-fish, as practiced in the Greenland fishery, the *Caerwent* would clearly be entitled to possession of the whale as it was not attached to the *William Fennings* and was, therefore, a loose fish. Testimony, however, was adduced by the owners of the *William Fennings* that in the Southern fishery – of which the area around the Galapagos Islands was a part – a different custom prevailed. Since its inception, the Southern fishery awarded one half of the whale to the party that affixed a drogue and one half to the whaler that killed the animal. Counsel for the *Caerwent's* owners acknowledged that this had once been the custom of the Galapagos, but argued that since 1792 the custom of most captains was to award the entire whale to the ship that actually captured it. It was further shown that the captain of the *Caerwent* was one of six masters who had agreed upon his arrival in the Galapagos in 1805 to follow the emerging custom that eschewed the sharing of whales. Chief Judge Mansfield determined that as the captain of the *William Fennings* was not a party to that agreement, the Southern fishery custom of dividing the whale must be followed. Judge Chambre advanced in his concurrence the importance of following custom in an area of commerce engaged in by the subjects of many nations. Failure to abide by established customs would result in a sort of warfare between ships that might eventually extend to the nations of those involved.[22]

While much of the discussion of the judges in *Fennings* concerned whether the action was properly brought in trover, the arguments of counsel illuminated the gap between custom in theory and in how it was used to shape English substantive law. Counsel for the *Caerwent* argued that the practice of splitting whales in the Southern fishery did not constitute "a custom of a particular trade as to be binding in law" or, in the alternative, that if it was a custom, it was no longer followed. Given the apparent disagreement between captains plying these waters as to established practice, the *Caerwent* suggested that the common law rule of *ferae naturae* be applied. Not surprisingly, the *Caerwent* – as a second striker – asserted that the common law honored the claim of

[22] *Fennings v. Lord Grenville*, 1 Taunt. 241. Drogues used in the Southern fishery commonly consisted of pieces of wood attached to the lines. Granville Allen Mawer, *Ahab's Trade: The Saga of South Seas Whaling*, (New York: St. Martin's Press, 1999), 350. The agreement referenced in *Fennings* has not survived, but for an example of an agreement by Galapagos whalemen from many nations providing that the first striker was not entitled to a portion of the whale from the vessel that ultimately dealt the fatal blow, see Edouard A. Stackpole, *Whales & Destiny: The Rivalry between America, France, and Britain for Control of the Southern Whale Fishery, 1785–1825* (Amherst: University of Massachusetts Press, 1972), 388–9.

the party that gained actual physical possession of the prey. Given the strict definition of custom as being synonymous with common law, a reader might be excused for wondering what counsel meant when he suggested that, in the case of the Southern fishery, a custom was not always binding. Furthermore, if *ferae naturae* was the common law and, therefore, custom, why were other practices even under consideration?[23]

The reason for this confusion can be found in the difficult position imposed on English judges and lawyers by strictly declaring the common law to be custom. How did the common law evolve to accommodate new problems and rethink old issues if courts were bound to follow practices that date back into the misty English past? The solution was to pay homage to the theory that there was no such thing as a new general custom, while recognizing the existence of fresh, limited customs that were subject to a different set of rules. A local custom was one that supplanted a general custom in a particular region through observance from time immemorial or, at least, for a long time. The theory appeared to be that such local customs shared with general customs the same hoary past that demanded recognition as law. Unlike customs general throughout the realm, it was possible for courts to recognize local customs that, despite their antiquity, had not previously come to the attention of courts or legal scholars. The acknowledgment of local custom was not by itself, however, a sufficient mechanism for change. What status should be given to the often dynamic and ever evolving practices observed in particular trades or businesses that were essential to the British economy? Such recent norms and practices were termed usages and while denied the power of custom to change law, they were allowed to be read into and interpret contracts between participants in a particular trade so long as they did not run counter to established law. The recognition of local custom and usage was useful, but did not allow courts the desired room for innovation.[24]

The ultimate solution was a blending of the concepts of general custom, local custom, and usage. Courts grew accustomed long before *Fennings* to permitting even usages to create law. Counsel for the *Lord Grenville* reflected a degree of

[23] *Fennings v. Lord Grenville*, 1 Taunt. 242. Despite the court's determination that the existing custom favoring the *William Fennings* was applicable, a nonsuit was entered dismissing the action brought by its owners. Pursuant to English law, the owners of the two ships were deemed tenants in common in the whale. Tenants in common were often said to have a unity of possession. Joint tenants enjoy an equal right to possess the entire property held in common. The mistake made by the owners of the *William Fennings* was to bring this action in trover. An action in trover between tenants in common was only permitted when the common property had been destroyed. It was determined that killing a whale at sea and rendering its remains into oil constitutes preservation, rather than destruction. In other fisheries where the custom did not call for the sharing of a whale, trover was the preferred cause of action.

[24] Cameron, "Custom as a Source of Law in Scotland," 306–21. For an example of how an English court viewed a recently developed usage in a particular business as altering the terms of an insurance policy, see *Noble v. Kennoway*, 2 Dougl. 510, 99 Eng. Rep. 326 (1780).

frustration with the traditional common law theory as to how laws come into being with the observation – supported in true English fashion with applicable precedent – that in resolving the case at bar it was immaterial whether the practice in the Southern fishery "be called an universal agreement in the trade, or an usage, or a custom." The important issue was whether "this practice as in the nature of a law, . . . possesses the quality so essential to that character, of being highly reasonable." Judge Chambre's opinion shared much of counsel's pragmatism in his concern that whatever rule was adopted it best be one that prevented the international cast of participants in the Southern whalefishery from sparking warfare at sea. "The greatest of all legal fictions," E. P. Thompson has concluded, "is that the law evolves, from case to case, by its own impartial logic, true only to its own integrity, unswayed by expedient considerations." To Thompson's observation might be added that the core explanation for how law is created in English common law is also a fiction motivated by a healthy dose of expediency.[25]

IV

The arguments of counsel before the Court of Session and President Judge Campbell's ruling in *Addison v. Row* illustrate how differently Scots law viewed custom as a mechanism for change. Counsel for both ships presented an argument straight out of Justinian's *Digest* that would have been immediately recognizable to anyone who had studied Roman law since its revival in eleventh-century Bologna. Drawn from Book 41, title 1 of the *Digest*, the issue was at what point in the pursuit of a wild animal was a first striker deemed to be in possession of its quarry.[26]

Counsel for the *Caledonia* asserted that fast-fish, loose-fish was the custom of the fishery and should be honored because it accorded with the general principles of the Roman law concept of *ferae naturae* that gave ownership to the party that gained actual possession of the beast. The *Priscilla* countered that the requirement of physical control was an antiquated position, suitable to a more primitive notion of property rights. The modern rule, favored by more recent Roman lawyers, credited the first striker and was based on the advanced notion that the possession of property was more properly an act of the mind than a race to hold an animal. President Judge Campbell, in explaining his decision, first considered which view of possession represented "the general rule." Citing from the *Digest*, Campbell agreed with the *Priscilla* that the recent trend was to loosen the degree of possession needed to secure a wild animal. "There is no custom proved," Campbell continued, "which can derogate from this general principle."[27]

[25] *Fennings v. Lord Grenville*, 1 Taunt. 244. E. P. Thompson, *Whigs and Hunters: The Origin of the Black Act* (New York: Pantheon Books, 1975), 250.

[26] *Addison v. Row*, in *Reports of Cases Decided in the House of Lords*, 3:337,344.

[27] *Addison v. Row*, in *Reports of Cases Decided in the House of Lords*, 3:337–8.

The reader is, thereby, left to wonder whether Campbell was saying that the *Caledonia* failed to demonstrate that fast-fish, loose-fish was the custom of Greenland or that no custom was competent to overcome a general principle of Roman law. The likely answer can be found in the Scottish approach to custom. Erskine explained, following the Roman model, that Scots laws were either written or unwritten. Written law as set forth in statutes clearly carried the express authority of the duly constituted officials. The unwritten law of Scotland was customary and derived its binding status from the tacit approval of the legislature. This customary law – also referred to as the common law of Scotland – consisted of "ancient usages, . . . whether derived from the Roman law, the feudal customs, or whatever other source."[28]

Erskine indicated that some usages, such as primogeniture or the widow's terce, were so ancient and accepted that no proof of their status was necessary. Other customs, of a more recent accretion, required evidence of the "antiquity and universality" of their usage. When Scots lawyers spoke of a custom being of immemorial usage, however, they did not, like their English brethren, imagine a practice so old that its origins could not be discerned. Instead, a custom in Scots law could slowly gain acceptance, requiring no particular length of existence or number of usages to be deemed valid. As Erskine explained, "some things require in their nature longer time, and a greater frequency of acts, to establish them, than others."[29]

While the absence of a prescribed period of time before a practice could become a custom might seem to indicate an easier path to establishment in Scotland, the opposite was true. Recall that in England only a local custom could create a new law. All general customs were already part of the common law. If a local custom was clearly at odds with the universal custom, an English legal theorist was not troubled, as both practices had been blessed with the imprimatur of immemorial usage. Scots law, permitting both universal and local customs to evolve slowly, required that its customs be tested for accordance with the core principles of the law. Antiquity could not save a custom that violated the accepted common law of Scotland which, as we have seen, was heavily dependent upon Roman Law.[30]

The differences between Scottish and English practice in this regard can be seen in the latter's embrace of the principle – set forth in the maxim *communis error facit jus* – that a long-held mistake as to the substance of the common law should be permitted to stand even after its discovery. The Scots, pledged to a common law that best reflected the aspirations of society, would quickly act to correct any such errors. The practical result of this distinction was that Scots jurists such as Campbell were reluctant to look to customs or usages for guidance. Why risk adopting a new way of looking at a problem when the

[28] Erskine, *An Institute of the Law of Scotland,* 1:10.

[29] Erskine, *An Institute of the Law of Scotland,* 1:16.

[30] Cameron, "Custom as a Source of Law in Scotland," 306–13.

situation was already covered by Roman law? Despite the theoretical acceptance of custom as a source of new law, there exists, as legal historian J. T. Cameron has remarked, "no single clear case of a customary rule being accepted as law by a Scottish court."[31]

Accordingly, when Judge Campbell was faced with the facts in *Addison v. Row*, he turned to the well-established Roman law discussion of *ferae naturae*, not the norms of the whaling industry. It was, of course, true that fast-fish, loose-fish, and iron holds the whale embodied the competing arguments set forth in Justinian's *Institutes*. Although Judge Campbell undoubtedly recognized the similarities, his inclination was not to find a new custom, but to determine which rule of whaling was, in principle, the most just. That he chose to use the language of the Roman law and not that of immemorial usage is not surprising. Confronted with the exact same situation, although framed by counsel in accordance with English common law, the House of Lords sought to determine how whalemen resolved such arguments. For the House of Lords, the stakes were not particularly high. Whaling was a very particular industry. Despite its importance to the British economy, a whaling practice given legal approval would have little role in shaping the law in other areas of commerce. To a Scottish judge the issue was not just how whalemen operated in a remote part of the world; it was about a higher principle of justice.[32]

Having been declared by the House of Lords to be the custom of Greenland whaling, fast-fish, loose-fish enjoyed a long run in British courts and legal treatises as the largely unchallenged law of whaling. The competing idea of iron holds the whale appears to have generally died out among Greenland whalemen after 1800 with the establishment of fast-fish, loose-fish as the legal standard. Yet, the idea that a first striker in pursuit deserved some consideration abided in the icy waters of the Greenland fishery. As Scoresby made clear, there was something unseemly about a first striker, in active pursuit, losing a whale to an opportunistic interloper.

V

On September 21, 1826 the *Old Middleton* spied a whale while standing off the coast of Greenland in company with the *Andrew Marvel* and the *Resolution*. A boat dispatched from the *Old Middleton* succeeded in affixing a harpoon and the whale – as was often the case – swam off at a fast rate of speed with the attached boat in tow. Later explaining that it believed the whale to be loose, the crew of the *Andrew Marvel* took up the chase, managing to strike and kill its target. Whether the *Andrew Marvel* had been in a position to observe the *Old*

[31] For a brief discussion of *communis error facit jus*, see Herbert Broom, *A Selection of Legal Maxims, Classified and Illustrated* (London: A. Maxwell & Son, 1845), 51. Cameron, "Custom as a Source of Law in Scotland," 313, 321.

[32] *Addison v. Row*, in *Reports of Cases Decided in the House of Lords*, 3:338–40.

Middleton's original assault is unclear. It is likely, in any event, that what began as an attempt to assist the *Old Middleton* changed when it became apparent to the *Andrew Marvel* that its competitor's line had come loose from the harpoon. In the London Court of Common Pleas, counsel for the *Andrew Marvel* cited *Fennings v. Lord Grenville* for the time honored proposition that a whale remained fast only so long as the connection between the first striker's boat and the affixed harpoon remained intact. The *Old Middleton* countered that the custom of fast-fish, loose-fish had evolved in the more than three decades since *Fennings* was decided. The evidence presented at trial proved to the court and forced the *Andrew Marvel* to concede that in the Greenland fishery a whale was now considered fast – even absent an attached harpoon – if it was entangled in the line which remained in control of the original striker. The jury found that the whale had remained fast and rendered a verdict for the *Old Middleton*.[33]

While the change in custom set forth in *Hogarth v. Jackson* – as this case was captioned – was slight and a reasonable accommodation for a common occurrence that did little to change the basic tenor of the practice, it reveals a fluidity in whaling customs not readily apparent in many scholarly treatises that continued to cite *Fennings* as the prevailing standard long after the British had abandoned the trade. The impetus for this incremental alteration in the custom was likely the feeling of most whalemen that making the continued attachment of the harpoon determinative was, as Scoresby would have agreed, unfair to the first striker. That English courts – always concerned with following the lead of whalemen in setting the customs in an international industry – would have been sensitive to changes at sea is not surprising. What is curious is the degree to which trial judges and lawyers by the 1820s seemed to be leading the whalemen, pushing the law from fast-fish, loose-fish to the standard of iron holds the whale. Although it may be that jurists were simply reacting to evidence adduced from whalemen that was not preserved in the reported decisions, the evidence from other trials in the same period does not indicate that whalemen were moving in this direction.[34]

33 For the various sources for *Hogarth v. Jackson*, see 2 Car. & P. 595, 172 Eng. Rep. 271; 1 Moo. & Malk. 58 (1827); *Times* (London), March 2, 1827, p. 3; and *Morning Chronicle* (London), March 2, 1827. It is not clear if the *Resolution* involved in this matter was the same ship captained by Scoresby in 1811 and 1812. Scoresby served as the master of other vessels beginning in 1813. Jackson, *The Arctic Whaling Journals of William Scoresby*, xxiii.

34 For post-*Hogarth* legal treatises that continued to state the law of whaling as set forth in *Fennings*, see, for example, George Lyon, *Elements of Scots Law in the Form of Question and Answer: With a Copious Appendix Containing Forms of Writings for the Purpose of Reference and Illustration* (Edinburgh: Thomas Clark, 1832), 3–4; and J. H. A. Macdonald, *A Practical Treatise on the Criminal Law of Scotland* (Edinburgh: W. Patterson, 1867), 24. For two other whaling cases from this period, see *Hutchison v. The Dundee Union Whale Fishing Company* reported at Alexander Peterkin, *Whale Fishery: Report of the Trial by Jury, John Hutchison, Esquire, and others, Against The Dundee Union Whale Fishing Company* (Peterhead, Scotland:

This fundamental uncertainty as to what precisely transpired at court was endemic to Anglo-American law in this period. Reports of cases were produced privately for profit and were not subject to court approval. Although the reporters were members of the bar, the quality of their work varied greatly. Only a small percentage of trials were reported and the criteria for inclusion in the published volumes were rarely made clear. One early nineteenth-century reporter, John Campbell, famously remarked in his autobiography that he suppressed decisions of Lord Ellenborough that he deemed "inconsistent with former decisions or recognised principles." Whether Lord Ellenborough should have been grateful – as Campbell bragged – is unclear as the "bad Ellenborough law" has not survived. The problem of selective and possibly incompetent reporting was not new. Blackstone lamented that since the early sixteenth-century private reporters "through haste and inaccuracy, sometimes through mistake and want of skill, have published very crude and imperfect (perhaps contradictory) accounts of one and the same determination." The reports produced for the trial of *Hogarth v. Jackson* illustrate this problem of interpreting competing reports that confronted contemporaries and has vexed later scholars.[35]

The discussion of *Hogarth* presented in *Carrington & Payne's Nisi Prius Reports* was brief and unequivocal. The custom of fast-fish, loose-fish set forth in *Fennings* had, it was explained, been changed so that a whale remained fast "whether the harpoon continues in the body or not, if the fish is attached by any means, such as the entanglement of the line, or other cause, to the boat of the party first striking it, so that such party may be said to have power over it [the line]. . . ." It was also reported that Chief Judge Best opined that this new custom represented an improvement in that it was not always easy to determine in the water whether a harpoon remained affixed. A perceptive reader of *The Times* blessed with a good memory might well have reviewed Frederick Augustus Carrington and Joseph Payne's report of the *Hogarth* trial with some confusion. *The Times* account, carried the day after the trial, explained that the custom had long been that when a whale is "struck by a boat, in such a manner that there shall be little doubt of the crew being ultimately able to kill it, it is considered the property of that boat." The custom of fast-fish, in *The Times'* telling, sounded very much like the law advanced by Judge Campbell in *Addison v. Row*. While it is tempting to dismiss *The Times* account as the

P. Buchan, 1830) and also at Joseph Murray, *Report of Cases Tried in the Jury Court, at Edinburgh, and on the Circuit, From November 1828 to July 1830, Both Inclusive* (Edinburgh: G. A. Douglas, 1831), 162–5; and *Nicoll v. Burstall* covered in *Times* (London), February 27, 1834; *Morning Chronicle* (London), February 27, 1834; *Hull Packet*, March 7, 1834; and *North Wales Chronicle* (Bangor, Wales), March 11, 1834.

35 For John Campbell's quotes and a discussion of trial court reports, see James Oldham, "Law-making at Nisi Prius in the Early 1800s," *Journal of legal History* 25, issue 3 (December 2004), 223. Blackstone, *Commentaries on the Laws of England*, 71.

product of a reporter unskilled in the ways of the law, the version of *Hogarth* contained in *Moody & Malkin's Nisi Prius Reports* suggests that the newspaper's coverage of the case – while misleading as to the custom of the fishery – was not entirely inaccurate.[36]

William Moody and Benjamin Heath Malkin framed the argument between the parties in different terms than that presented by their competitors, Carrington and Payne. In Carrington and Payne's account, the *Andrew Marvel* argued in favor of the *Fennings* custom, while the *Old Middleton* convinced the court that a new standard no longer required that the harpoon remain in the whale and attached to an entangled line. Moody and Malkin, on the other hand, indicated that the *Andrew Marvel* conceded that the harpoon's position was not determinative and that the dispute between barristers centered on the degree of control the first striker must retain over the line in which the whale was entangled. The *Old Middleton*, as first striker, maintained that a whale was fast if the rope remained in control of the boat and "was any how attached to the fish." The *Andrew Marvel* countered that a whale was not – pursuant to the new custom – entangled "unless she [the whale] were so fast in the rope as to give the first strikers the same power over her as if the harpoon continued fixed." The difference in how the competing reporters presented the argument may be the result of Carrington and Payne providing the initial arguments of counsel. As Moody and Malkin indicated, the *Andrew Marvel* at some point conceded that the custom had changed. Perhaps, it was at this point that the defendant's position changed to reflect the obviously compelling evidence of a new practice and came to be that presented by Moody and Malkin. The most interesting difference between the two reports, however, was in the role assigned to Chief Judge Best. In Carrington and Payne's account, Best simply gave his imprimatur to the change in custom. The judge, however, emerged in Moody and Malkin's telling as a full participant in shaping custom into law. Best expressed his view that the *Old Middleton* was fast if the custom was "understood to extend to all cases where the whale was so far *entangled* in the rope of the first strikers, that they might thereby have a reasonable expectation of securing her." (italics in the original) The judge appeared to have simply decided that the test should include an assessment of the first striker's prospects for capturing the whale.[37]

Given the concordance between the accounts of *The Times* and Moody and Malkin, it is highly likely that Chief Judge Best did, indeed, interject the idea that the likelihood of the first striker completing capture of a whale – absent intervention – was relevant to deciding such disputes. As reporters, Moody and Malkin appear to have not only recorded Judge Best's legal analysis, but also

[36] *Hogarth v. Jackson*, 2 Car. & P. 595; 1 Moo. & Malk. 58; and *Times* (London), March 2, 1827. *Morning Chronicle* (London), March 2, 1827, is alone among the accounts in viewing the case as being decided according to *Fennings*.

[37] *Hogarth v. Jackson*, 2 Car. & P. 595–7; and 1 Moo. & Malk. 58–60.

relaxed the initial boat's burden of proof. In the printed marginalia common in legal reports of this era, Moody and Malkin characterized the custom as entitling the first striker to a whale "though his harpoon be detached from the line when the second striker strikes, if the fish be so entangled in his line that he might probably have secured her without the interference of the second striker." Moody and Malkin's "might probably" would seem to require a lesser probability of success than the "reasonable expectation" ascribed directly to Judge Best. The role of Moody and Malkin in shaping – or at least attempting to shape – whaling law can also be seen in their decision to include as part of their *Hogarth* report a description of another matter involving a disputed whale.[38]

In *Skinner v. Chapman*, heard at the 1827 York Lent Assizes, a whale fast to the *Harmony's* harpoon and line was lanced by a boat belonging to the *Phoenix*. While the lance did nothing to assist in the capture of the whale, it did agitate the animal to the extent that it broke free of the harpoon. The *Phoenix* argued at trial that it subsequently harpooned what was then a loose target. The first striker countered that the second boat's harpoon was friendly in that it had been affixed prior to the moment when the whale's violent movements sundered the connection with the harpoon and line. Judge Bailey, as stated by Moody and Malkin, charged the jury that if they determined that the first striker's harpoon had already come loose when the second iron was affixed, they must then decide "whether the plaintiffs [first striker] could have secured the fish if the lance of the defendants had not been struck." To this instruction, Judge Bailey added that it was his belief that if a whale had been successfully struck and an unsolicited party "does an act which prevents the first striker from killing it, and then kills it himself, he kills it, not for his own benefit, but for that of the first striker."[39]

There is, however, no evidence that whalemen in Greenland were developing a custom in this period that included any such calculation of the first striker's likelihood of success. The arguments of counsel in the 1830 Scottish case *Hutchison v. The Dundee Union Whale Fishing Company* made, for example, no reference to the probability of capture even though *Hogarth* and *Skinner* were cited as precedent. The testimony of the witnesses in *Hutchison*, summarized in a 34 page pamphlet reporting the trial, makes clear that the questions of the advocates were not directed at the first striker's prospects absent the interference of another vessel. Similarly, in *Nicoll v. Burstall*, tried before the Court of Exchequer in 1834, the statement of the custom contained in *The*

[38] *Hogarth v. Jackson*, 2 Car. & P. 595–7; and 1 Moo. & Malk. 58–60.

[39] *Skinner v. Chapman*, 1 Moo. & Malk. 59. The account of the trial in *Leeds Mercury*, April 7, 1827 appears to be largely a recitation of the argument made by the *Harmony's* counsel. The newspaper does support Moody and Malkin's discussion of the finding of the jury; stating that the use of a lance to free the whale from the *Harmony's* harpoon was "a wrongful act" and that the *Harmony* would have taken the bowhead if the *Phoenix* had not interfered.

Times intimated that the rule in *Fennings* was still being followed in the Greenland fishery. "The rule of the whale fishery," reported *The Times*, "was, that if the crew of any ship fix the harpoon into the whale and fasten it by their line to their boat, the crews of other vessels are not allowed to interfere with that whale, excepting indeed with the view of offering assistance." While the *Nicoll* court pointed to the necessity of an affixed harpoon, it also added – reflecting, perhaps, the *Skinner* decision – that an intervener cannot gain from preventing a first striker from killing a whale to which it was fast.[40]

If whalemen plying the waters of the Greenland fishery were not particularly concerned with the first striker's probability of capturing a whale, why were the judges in *Hogarth* and *Skinner*? The key to resolving this question can be found in Carrington and Payne's account of *Hogarth*. Whereas Moody and Malkin indicated that the cause of action was in trover, Carrington and Payne revealed that, in addition to trover, "[t]he declaration charged the defendants with having interrupted the plaintiffs in killing a whale." To prevail in an action in trover, the plaintiff must prove "a property in himself" in the object of the dispute, "a right to the present or absolute possession of them," and that the defendant converted the property for his own use or refused a demand that the item be returned. Ownership or actual possession was not required. It was sufficient if the plaintiff had "a right of present possession, which the Plaintiff may immediately take if he pleases." Trover proved to be a convenient cause of action for whaling disputes where an attached first striker did not actually possess the whale, but had a present possessory interest that the law was willing to recognize.[41]

Trover was not, however, an action that could be maintained if the plaintiff's right was deemed to be one of future possession. In bringing the action in *Hogarth*, counsel for the *Old Middleton* likely reasoned that if the court failed to accept the new custom which honored an entangled whale absent an affixed harpoon as fast, his client would not prevail. Accordingly, the *Old Middleton* needed to state a cause of action that was not based on the level of possession required in trover or by the custom of the fishery. As Carrington and Payne reported, the *Old Middleton* also claimed that the *Andrew Marvel* had "interrupted the plaintiffs in killing a whale." What the *Old Middleton* had set forth

[40] *Hutchison v. The Dundee Union Whale Fishing Company*, Peterkin, *Whale Fishery*, 162–5. *Nicoll v. Burstall*, *Times* (London), February 27, 1834, p.4; *Morning Chronicle* (London), February 27, 1834; *Hull Packet*, March 7, 1834; and *North Wales Chronicle* (Bangor, Wales), March 11, 1834.

[41] *Hogarth v. Jackson*, 2 Car. & P. 595–7; and 1 Moo. & Malk. 58–60. Isaac 'Espinasse, *Practical Treatise on the Settling of Evidence for Trials at Nisi Prius; and on the Preparing and Arranging the Necessary Proofs* (London: Joseph Butterworth and Son, 1825), 418, 432. The requirement in trover that a plaintiff be in a position to immediately take possession of a disputed object does not seem to have ever been strictly enforced in whaling cases where a first striker's ability to take immediate possession was often questionable. The concept of fastness seems to have been used as conclusive evidence of a right to immediate possession.

was a claim based not on their present possession of the whale, but on their right to continue pursuit of the whale without the interference of another party. The *Old Middleton's* rights, pursuant to this cause of action, were limited. It had the right to chase the whale without obstruction until it either succeeded in killing the animal or it no longer was in a position where capture appeared likely. This alternative cause of action – what British courts would later designate the tort of interference with trade, profession or calling – was not well established in the 1820s. In the first decades of the twentieth century, legal scholars argued whether the tort was a late nineteenth-century development or one of more ancient standing. Without recounting the specifics of this scholarly tussle, which seems to have had more to do with contemporary British labor relations law than a conscientious dispute over the historical development of a cause of action, there were numerous cases available in the 1820s that at least suggested that a cause of action could be successfully brought to prevent interference with a plaintiff's trade. The most famous of these cases was the 1707 matter of *Keeble v. Hickeringill*.[42]

In *Keeble v. Hickeringill*, the plaintiff, Keeble, had a decoy pond on his property that attracted ducks for capture. The neighbor, defendant Hickeringill, twice fired guns from his own property seeking to scare away Keeble's ducks. Keeble brought an action alleging that Hickeringill had interfered with his ability to gain a profit from his decoy pond. Hickergill countered that Keeble did not own the ducks on his property and therefore did not have a cause of action. The court agreed that Keeble did not own the ducks, but pointed out that the action was not brought to recover property. Instead, the court held that Keeble had the right to enjoy the benefits of his decoy pond, free from the malicious interference of his neighbor. "He that hinders another in his trade or livelihood," Chief Judge Holt explained, "is liable to an action for so hindering him." The court hastened to add that Keeble's rights would not have prevented Hickergill from exercising his concomitant privilege of building a decoy pond on his own property. In the context of a whale capture dispute such as *Hogarth*, a first striker who was not fast to a whale, but had wounded his quarry, had a limited possessory interest in the whale so long as he maintained the chase with the prospect of success. Similarly, Keeble had a limited sort of possession in ducks while they were on his property.[43]

While *Skinner* was, according to Moody and Malkin, brought in trover, it is very likely that there was an additional count of interference with trade. If the

[42] *Hogarth v. Jackson*, 2 Car. & P. 595–7; and 1 Moo. & Malk. 58–60. For the dispute concerning the history of interference with trade and the relevant cases, see Sarat Chandra Basak, "Principles of Liability for Interference with Trade, Profession or Calling, *Law Quarterly Review* 27 (1911): 290–312; and Thomas Atkins Street, *The Foundations of Legal Liability: A Presentation of the Theory and Development of the Common Law* (Northport, NY: Edward Thompson Company, 1906), 1:347–64. *Keeble v. Hickeringill*, 3 Salk. 9, 91 Eng. Rep. 659 (1707).

[43] *Keeble v. Hickeringill*, 91 Eng. Rep. 659.

action was based solely on trover, Judge Bayley's instruction to the jury that they consider the likelihood of the first striker securing the whale absent the interloper's lance thrust would not be legally relevant. In trover, the first striker's possession and, therefore, legal right to the whale was conclusively established by the connection from the fish to the boat. Once this is established, the actual prospects of capturing the whale were irrelevant. It did not matter, for example, that at the moment after the intervener sank its harpoon into a whale the first striker's boat lost its line in a spectacular collision with ice that all witnesses agreed could be foreseen moments before its actual occurrence. The whale, in these circumstances, belonged to the first striker and the second boat's harpoon was deemed friendly.

The likelihood of capture was interjected by attorneys and discussed by judges in *Hogarth* and *Skinner* because it was relevant to another cause of action governed by principles different from those at work in fast-fish, loose-fish. Although British whalemen in Greenland never abandoned the underlying principles of fast-fish, loose-fish, the idea that the prospects of a boat's ultimate success was worthy of some consideration was not entirely antithetical to the men who practiced the whaling trade. Scoresby's belief that his father was entitled to the bowhead taken by the *Resolution* in the summer of 1812 reflected the same notion of fairness at work in the tort of interference with trade. It also explains the numerous times captains agreed among themselves at sea, without strict adherence to custom or law, as to the fairest distribution of a contested whale. Scoresby called it the "laws of honor" and Melville explained that "upright and honorable whalemen" made allowances in situations where employment of custom would constitute "an outrageous moral injustice" as applied to a worthy party previously in pursuit of a whale taken by another vessel. Whalemen in Greenland made custom that British courts adopted and fit into existing causes of action. Creative British lawyers and courts also looked at whaling disputes and found that desired results could be reached through use of a different cause of action. Whalemen likely never made any real adjustments in their practices to reflect the rulings in *Hogarth* and *Skinner* because Scoresby's "laws of honor" were already well established in Greenland. It would be for American courts to establish – or reestablish if seventeenth- and eighteenth-century practices are considered – pursuit of a detached wounded whale with the prospect of success as the standard that came to be known as iron holds the whale.[44]

The creation of British whaling law was not – as some legal scholars might frame the question – either imposed from above or crafted by participants. It was a combination. Whalemen developed a custom that fit well with the

[44] Jackson, *The Arctic Whaling Journals of William Scoresby*, 119. Melville, *Moby-Dick*, 433. Benjamin L. Fine, "Analysis of the Formation of Property Rights Underlying Tortious Interference with Contracts and Other Economic Relations," *University of Chicago Law Review* 50 (Summer 1983), 1116–45.

existing Roman and common law concepts of how wild animals come to be owned. The men who plied the whaling trade thought about property in the same manner as their contemporaries. Property rights, they believed as Locke explained, are created by an application of effort to an object that was not previously owned. Whaling disputes and the customs and laws established for their resolution were really just arguments about when a whaleman has done enough to earn the right to claim a whale. Competing positions were points on a continuum from lowering boats upon sighting a whale to, at the opposite end, the physical possession of a dead whale. Having created these norms, Greenland whalemen never hesitated to seek resolutions to particular disputes that ignored law and customs; favoring instead mutual agreements that likely were based, in part, on factors such as the history between ships and the group dynamics of numerous vessels sailing in company over the course of months. British courts, in turn, honored whaling customs, while, at the same time, changing the law as part of the ongoing nineteenth-century evolution in the available causes of action.

The British whaling cases offer evidence to both the legal centralists and the norms scholars in their efforts to explain how property law is created. The decisions of British courts recognizing fast-fish, loose-fish over iron holds the whale by the 1780s appears to be a clear example of a court imposing its will on involved parties. Yet, as Scoresby's discussion of whaling law indicates, fast-fish, loose-fish was widely accepted by whalemen in the Greenland fishery even prior to the involvement of British courts. Iron holds the whale – while clearly an alternative custom followed by some whalemen before 1800 – was likely a minority practice that might well have disappeared even absent the death sentence imposed by British courts. The creation of British whaling law was, in fact, largely a collaborative effort of whalemen, judges, and lawyers. The end result by the 1780s was a blending of custom and law that whalemen followed and courts honored. British whaling law and custom also retained a certain degree of fluidity as whalemen continued to abide by their laws of honor and courts were certainly willing – as they did in the 1820s – to alter the law.

4

Iron holds the whale

American whaling law, as explained in early nineteenth-century legal treatises directed at attorneys, was indistinguishable from British practice. In the absence of reported American cases, American lawyers and judges were offered the guidance of British decisions. In the eighteenth century, whaling property disputes were litigated in American courts and were the subject of legislation in the seventeenth century. The failure of American judicial opinions from the colonial period to be preserved in printed form had, however, effectively erased these decisions from the legal memory upon which the common law is built. In turning to British cases, American attorneys and legal scholars took comfort in the apparent stability of the customs and laws that had governed whaling since the dawn of the trade. That this stability was, in part, imposed by judges and practitioners in London was not, of course, recognized in America. In the effort of American judges and lawyers to find widely accepted customs and create a law as clear and as stable as that of the British, they followed their colleagues from across the Atlantic in failing to recognize the fluidity and improvisational ways in which whalemen – British and American – actually resolved their disputes.

Nineteenth-century American legal scholars largely failed, in their reliance on British cases, to see that American whalemen had a long tradition of iron holds the whale and not – like their British colleagues – fast-fish, loose-fish. Later legal and whaling historians, familiar with the British decisions and Herman Melville's insistence that fast-fish, loose-fish was practiced by all whalemen, have also failed to recognize that beginning at least in the eighteenth-century American whalemen generally followed the custom of iron holds the whale. The tenacity of this misunderstanding of American practice can still be seen in the belief that American whalemen only switched to iron holds the whale once they switched to hunting bowheads and other right whales in the 1840s.[1]

I

When American Joseph Angell explained ownership rights in fish found in the ocean and other navigable waters in his 1847 edition of *A Treatise on the Right of Property in Tide Waters and in the Soil and Shores Thereof*, he employed a mixture of Roman law, common law, and custom drawn almost exclusively from British sources. That Angell – one of a group of American legal scholars writing in what Morton Horwitz has dubbed the early nineteenth-century "treatise tradition" – should cite Bracton, Fleta, and British cases such as *Hogarth*, reflects the degree to which English common law continued to provide the foundation for antebellum American law. Like James Kent, Joseph Story and the other American legal treatise writers, Angell was interested in providing reliable, comprehensive guides to subjects of interest to practicing attorneys. The antebellum treatise writers were part of a conscious effort to supplant an earlier generation of legal thinkers who were convinced that natural law and the study of political philosophy were the surest guarantors of American liberty and prosperity. For Joseph Angell and his colleagues, law – properly understood – was a science. The employment of scientific principles of classification and deductive reasoning would impose a new objective rigor that would free law from the politics and policy making that afflicted previous legal regimes. As legal historian Robert W. Gordon has observed, the treatise writers were painfully aware that the received legal wisdom upon which they hoped to build their new edifice was, in fact, the same English law that embodied many of the ills they sought to escape.[2]

To accomplish this difficult task of gleaning wheat from legal chaff, Joseph Story argued that a learned and independent judiciary was essential. A judge dependent upon the continued beneficence of a political patron or the goodwill of an untutored populace was not free to develop an American law that would "flourish as a science." Story – simultaneously a Supreme Court justice, treatise writer, law professor, and bank president – offered England's Lord Mansfield

[1] Robert C. Ellickson, *Order without Law: How Neighbors Settle Disputes* (Cambridge: Harvard University Press, 1991), 191–206.

[2] Joseph Kinnicut Angell, *A Treatise on the Right of Property in Tide Waters and in the Soil and Shores Thereof* (Boston: C. C. Little and J. Brown, 1847), 137–8. Morton J. Horwitz, *The Transformation of American Law, 1780–1860* (Cambridge: Harvard University Press, 1981), 253–66. Robert W. Gordon, "Historicism in Legal Scholarship," *Yale Law Journal* 90, no. 5 (1981): 1038–9. Citing Lord Mansfield with approval, Angell declared "that the predominating characteristic of the Common Law is, that it is a *science of principles*." *A Treatise on the Law of Watercourses with an Appendix Containing Statutes of Flowing, and Forms of Declarations* (Boston: Charles C. Little and James Brown, 1850), iv (italics in original). For *Hogarth*, see Chapter 3.

and his incorporation of maritime and commercial law into the common law as an example of how law when treated as a science leads beyond uniformity in a single nation to a universally applicable law of nations. While acknowledging that commercial and maritime law by virtue of its cast of international participants has a tendency to assimilation, Story celebrated that the Atlantic nations had already achieved a notable confluence in their usages and statutory provisions. Story's prescription for how this might be accomplished in other areas of the law or, for that matter, precisely how commercial law reached this exalted state, was, alas, unclear. It appeared that this golden age would be achieved by simply permitting highly intelligent legal professionals to study multiple sources from the great legal texts to local customs and usages and thereby distill a sort of wisdom of the ages. Story, to be sure, had at least two clear agendas he sought to advance in expressing these theories at the 1821 annual Boston gathering of the Suffolk County bar. A federalist, like Angell and most of the treatise writers, Story advocated a powerful, independent federal judiciary competent to pass judgment on matters of constitutionality and capable of providing an example of legal science to state and local bars that might otherwise descend into a stultifying provincialism. Story also called for the gradual codification of different areas of law as they became the subject of sufficient scientific scrutiny. This would provide lawyers, legislators, and laymen with the tools to readily understand the correct principles of the law and thereby escape the ever growing morass of flawed, inconsistent opinions.[3]

Measured by the standard set by Story for a scientific explication of the law that would serve as foundation for future codification, Angell's discussion of how property rights in fish and whales were secured was decidedly lacking. Angell began with the observation that when such creatures are common to all, "it follows that an actual appropriation or manucaption must be made" to gain possession. No explanation other than perhaps the authority of Bracton and Fleta as oracles of the common law was offered to clarify why complete possession should be the *sine qua non* of ownership. Eliding the split in Roman law on this point and the view of Trebatius that possession of a wild animal

[3] Joseph Story, "An Address Delivered before the Members of the Suffolk Bar, at Their Anniversary, on the Fourth of September, 1821, at Boston," *The American Jurist and Law Magazine* I, no. 1 (January, 1829): 11. For additional discussion of the "treatise tradition," codification, and the call for a more scientific approach to the law in the early nineteenth century, see, in addition to the aforementioned, Howard Schweber, "The 'Science' of Legal Science: The Model of Natural Sciences in Nineteenth-Century American Legal Education," *Law and History Review* 17 (Fall, 1999): 421–66; John H. Langbein, "Chancellor Kent and the History of Legal Literature," *Columbia Law Review* 93 (April, 1993): 547–94; Charles A. Bane, "From Holt and Mansfield to Story to Llewellyn and Mentschikoff: The Progressive Development of Commercial Law," *University of Miami Law Review* 37 (1983): 351–77; Maxwell Bloomfield, *American Lawyers in a Changing Society, 1776–1876* (Cambridge: Harvard University Press, 1976), 58–90; and Perry Miller, *The Life of the Mind in America from the Revolution to the Civil War* (New York: Harcourt, Brace & World, 1965), 99–265.

vested with the infliction of a wound and a continued pursuit that rendered the creature's prospects for escape unlikely, Angell moved on to the 1844 English case of *Young v. Hichens*. In *Young v. Hichens*, the plaintiff cast a seine around a shoal of mackerel. The fish were surrounded by the net with the exception of a seven to ten fathom opening at which was stationed two of the plaintiff's boats to rile the water and discourage escape. A boat belonging to the defendant took advantage of the opening, entered the enclosure, and successfully cast its own seine within that of its competitor. Arguing that he was in possession of the mackerel and would have, but for the defendant's actions, delivered the catch safely to shore, the plaintiff brought an action in trespass to recover the value of the fish. The court acknowledged that the plaintiff was on the verge of taking possession of the mackerel and that, absent the defendant's interference, "it was in the highest degree probable" that capture would have been completed. This was not, however, sufficient to constitute the requisite degree of "custodia or occupation" to prevail at trespass. The plaintiff's actual power over the fish must, instead, be shown.[4]

Without any transition or even a paragraph break, Angell – citing *Hogarth* – proclaimed that the custom in the Greenland fishery honors a whaler whose line so entangles his prey "that he might probably have secured the whale without the interference of the second striker." Angell, likely seeking to provide the most current British cases in the absence of relevant American decisions, summarized *Skinner* and included as a note *Hogarth's* discussion of that matter. Missing from Angell's discussion was any context that would allow the reader to make sense of the role of custom in altering the common law of acquiring ownership of wild animals. An American lawyer seeking guidance on the disposition of a captured whale or other creature – Angell's intended audience – would certainly want to know if the Greenland custom applied in other fisheries. Furthermore, did American whalemen and those from other nations join their British colleagues in recognizing this custom? Although he was undoubtedly aware of *Fennings* and the fact that the practice in the Southern fishery was different, Angell chose not to explore the implications of competing local customs. Angell, like the authors of other American legal texts was, for the most part, content to recycle British sources with little curiousity as to actual whaling practices. It was not until a series of whaling disputes before American courts resulted in published opinions in the period from 1856 to 1872, that the standard distillations of American law took notice that whaling law was not uniform across the English speaking world.[5]

4 Angell, *A Treatise on the Right of Property in Tide Waters*, 137–8. *Young v. Hichens*, 1 D. & M. 598 and 115 Eng. Rep. 228 (1844). The *Hichens* court did opine that plaintiff might have succeeded had he brought a different cause of action. While the court did not explain which cause of action might have brought the plaintiff a favorable result, it is likely that they had an interference tort in mind.

5 Angell, *A Treatise on the Right of Property in Tide Waters*, 137–9. For *Skinner and Fennings*, see Chapter 3.

The paucity of American sources in legal treatises does not mean, as one might surmise, that whale capture disputes had never been litigated in America. The decisions in such cases were simply not available to practitioners, judges, or treatise writers. Reports of American cases were not published prior to the Revolution. Some accounts of cases handled by colonial courts in a particular locality were passed in manuscript form between local practitioners, but American lawyers well into the nineteenth century relied heavily on English reports and treatises in making their arguments. Opinions from state courts began to be published in the late 1780s and early 1790s, but lawyers in New York and Massachusetts waited until 1804 for printed case reports from their highest courts. With Joseph Angell's 1847 commencement of reports from Rhode Island, cases from every state were finally available. The persistence of English case authority was not simply a matter of necessity in a new nation slowly creating its own common law. While English statutory law could be easily enough discarded as the legacy of a colonial regime, the common law dating back generations across the Atlantic embodied a special inheritance treasured as the source of wisdom that protected the rights, liberties, and property of men. English common law, as Kent explained quoting Sir Matthew Hale, does not develop quickly as it is "not the product of the wisdom of some one man, or society of men, in any one age; but of the wisdom, counsel, experience, and observation, of many ages of wise and observing men." America in the first decades of the nineteenth century was viewed by Kent and his contemporaries as engaging in the slow process of creating a common law suited to "the genius of our institutions." Although some states explicitly incorporated only the English common law as it existed in 1776, subsequent British reports were in the first decades of the nineteenth century scoured for guidance in matters before American courts.[6]

II

Conflict in America over the ownership of whales dates almost to the beginning of the English settlement of Massachusetts. Whaling in the first decades after 1620 was a passive enterprise. Those wishing to reap the financial benefits of a rendered whale had to wait for a carcass to drift ashore. As Eric Jay Dolin has observed, the clearest indication that drift whales were a commodity of value and – it should be added – the subject of considerable rancor is that local

[6] For the development of American case reports, see Lawrence M. Friedman, *A History of American Law* (New York: Simon and Schuster, 1973), 282–92 and William E. Nelson, *Americanization of the Common Law: The Impact of Legal Change on Massachusetts Society, 1760–1830* (Cambridge: Harvard University Press, 1975), 168. For the reception of English common law, see Horwitz, *The Transformation of American Law*, 4–9; and Nelson, *Americanization of the Common Law*, 8–9. James Kent, *Commentaries on American Law* (New York: O. Halsted, 1826), 1:439–40, 455.

authorities imposed regulations governing their disposition. In crafting these rules, officials in Massachusetts and the English settlements on Long Island viewed drift whales in much the same way as the common law. Bracton observed in the thirteenth century that the crown enjoyed a special privilege in whales that superseded the ordinary customs controlling the possession of wild animals. The king, "as the minister and vicar of God," holds power over all of his subjects and, therefore, is entitled to possess things "which are annexed to justice and to peace." Whales and sturgeon, termed royal fish, fell into this category and could not "be separated from the crown, since they constitute the crown itself." This royal privilege extended, however, only to whales found on shore or those at such a short distance from land that the inevitability of their landfall was certain. While Bracton did not make clear the precise rationale for connecting whales with justice or peace, the royal privilege was, according to Blackstone, still honored in the eighteenth century. Blackstone provided, however, a clearer explanation of the royal interest in whales. In return for the crown's role in protecting the seas from pirates, the king by statutory authority owned all drift whales "on account of their superior excellence." The excellence of whales was, of course, related to the value of their oil and bone.[7]

The crown and its officials being at a great distance, American colonists in the seventeenth century made other provisions for drift whales. Plymouth specified that the proceeds from drift whales be equally divided between the colonial government, the town where the animal was found, and the finder. The Cape Cod town of Eastham declared in 1662 that a portion of each whale be allotted to support the minister. In 1644, the General Court directed that in Southampton, Long Island particular individuals on a rotating basis process any drift whales. In return for this service, a double share of the proceeds was awarded. A single share was given to "every Inhabitant with his child or servant that is above sixteen years of age." Massachusetts, perhaps wishing to separately consider the grant of each whale, decreed in 1649 that "any Whale, or such like great fish cast upon any shore, shall be safely kept or improved where it cannot be kept, by the town or other proprietor of the land, til the General Court shall set Order for the same." Despite these diverse resolutions, the colonists remained

[7] Eric Jay Dolin, *Leviathan: The History of Whaling in America* (New York: W. W. Norton & Company, 2007), 41–2. Henry Bracton, *De Legibus et Conuetudinibus Angliae*, Travers Twiss, translator (London: Longman & Co., 1878), 1:441, 443. St. George Tucker, *Blackstone's Commentaries: With Notes of Reference, to the Constitution and Laws, of the Federal Government of the United States; and of the Commonwealth of Virginia*, 5 volumes (Philadelphia: William Young Birch and Abraham Small, 1803), 2:290. William Cruise, *A Digest of the Laws of England Respecting Real Property* (London: Joseph Butterworth and Son, 1818), 270. Dale D. Goble, "Three Cases/Four Tales: Commons, Capture, The Public Trust, and Property in Land," *Environmental Law* 35 (2005): 824–9.

true to the English notion that whales were creatures of "superior excellence" that belonged to – if not the crown – the larger community.[8]

With the inevitable development of shore whaling by the middle of the seventeenth century came an increased number of drift whales. Whales that were harpooned off the New England coast frequently escaped before succumbing to their wounds and finding their way to the beach. A new claimant was thereby introduced to disputes in addition to the usual cast of contestants. In 1688, Plymouth ordered that all harpoons and lances be marked so that the boat that killed the whale could be readily identified. Thereafter, drift whales bearing a marked harpoon were often awarded to the successful whalemen, subject to the finder's claim of salvage. Those who discovered whales on the beach were not, however, always willing to accede to the claims of the whaleman who struck the fatal blow. Wishing to discourage disputes and secure evidence should litigation ensue, the General Court at Plymouth directed in 1690 that individuals be appointed to view drift whales and make a careful record of the animal's wounds and the particulars of its discovery. Fines and a loss of any right to profit that were levied against finders who, prior to official inspection, defaced a drift whale by "cutting, stabbing, or launcing" were likely directed at efforts to obscure evidence that the creature had been wounded at sea. Whalemen were also directed to immediately report any whale that was wounded or killed at sea and the nature of the injuries inflicted.[9]

Typical of the disputes between whalemen and finders was the 1718 matter of *Griffin v. Thomas* heard before the Court of Admiralty at Boston. Samuel Griffin and the crew of his boat were at sea about a league from Cape Ann where they mortally wounded a whale. "Stormy Winds" and the approach of night prevented further pursuit, allowing the whale to escape bearing Griffin's harpoon, rope, drogue, and buoy. Thirteen days later the whale's carcass was discovered by Gideon Thomas and two other men on a Plymouth County beach. When Thomas reneged on an agreement that he render the whale on Griffin's behalf, the latter sought a judgment asking for the bone and oil subject to the finder's usual right of salvage. A hearing date was set for a little more than two months after the contested events and an advertisement was placed in the local

[8] Dolin, *Leviathan*, 41–62; Elmo Paul Hohman, *The American Whaleman: A Study of Life and Labor in the Whaling Industry* (New York: Longmans, Green and Co. 1928), 24–5; Alexander Starbuck, *History of the American Whale Fishery from Its Earliest Inception to the Year 1876* (New York: Argosy-Antiquarian Ltd., 1964), 1:6–9; and *The book of the General Lawes and Libertyes Concerning the Inhabitants of the Massachusets, collected out of the records of the General Court, for the several years wherin they were made and established. And now revised by the same Court, and disposed into an alphabetical order, and published by the same authority in the General Court holden at Boston, in May 1649* (Cambridge, Massachusetts: Samuel Green, 1660), 83. Martha Bockée Flint, *Early Long Island: A Colonial Study* (New York: G. P. Putnam's Sons, 1896), 232.

[9] Dolin, *Leviathan*, 41–62. Nathaniel B. Shurtleff, ed., *Records of the Colony of New Plymouth in New England* (Boston: William White, 1856), 6:251–3.

newspaper informing potential claimants about the action. The court ruled in Griffin's favor determining that he had, indeed, inflicted the fatal wounds as evidenced by his marked iron. Complicating the proceedings was the appearance of an attorney on behalf of the Vice-Admiral seeking to protect the King's possible right to the whale. The argument of Griffin's counsel that the perquisite of Admiralty in the whale only applied in the absence of a valid claim apparently prevailed. While the intervention of the crown failed in this matter, other attempts by governors of Massachusetts and New York to claim drift whales under the ancient doctrine of "Royal Fish" prevailed. Particularly galling to whalemen were the efforts of Lord Cornbury as governor of New York in 1702 to extend the crown's traditional interest in drift whales to those caught at sea.[10]

The advent of shore and pelagic whaling by Americans also gave rise to disputes between whalemen competing at sea for the same whale. In *Davis v. Sturges*, Seth Davis and several other Barnstable men were hunting whales not far off the coast of their hometown, when, on January 9, 1720, they struck and fastened to a large whale. Their quarry proved powerful and combative. Despite remaining fast to the whale by way of an affixed harpoon and a controlled line, Davis obviously felt that their hold on the whale was tenuous. To guarantee capture, he called to the competing whaleboat of Thomas Sturges for assistance. The price of such aid was a 1/8 share of the whale. The negotiations, if any discussion was even necessary, were certainly brief. Sturges, also hailing from Barnstable, would have been well aware of the customary rate for assisting another boat in taking a whale. The common practice between two whaleboats of calling for and providing assistance took an unusual turn when the arm of Samuel Lambert became entangled in the attached line. The rope, likely in an effort to free Lambert, was allowed to slip loose of the boat. Davis briefly continued in pursuit of the whale before deciding to seek medical attention on shore for his crew member. The departure of Davis' boat left Sturges with a decision. Was the 1/8 share worth Sturges' effort in continuing pursuit of a feisty whale which he had not struck prior to Lambert's injury? The condition of the whale at this time is unclear. Davis' pleadings described the whale as "in her Agonies & Struggling with your Proponts" prior to calling for help. Yet, the whale was still lively enough that Sturges – having decided to continue the chase – called for the help of two additional boats each having been promised a 1/8 share of the catch. Davis, having deposited Lambert on shore, rejoined the hunt. By the time Davis fell into company with Sturges, the

[10] *Griffin v. Thomas* in John Noble, *A Few Notes on Admiralty Jurisdiction in the Colony and in the Province of the Massachusetts Bay* (Cambridge, Massachusetts: John Wilson and Son, 1905), 25–8. Dolin, *Leviathan*, 41–62, 395–6. Starbuck, *History of the American Whale Fishery*, 24–36. E. B. O'Callaghan, ed., *The Documentary History of the State of New York* (Albany: Weed, Parsons & Co., 1850), 3:372–84. For an advertisement concerning a hearing on a drift whale in addition to that provided by Dolin, see *Boston Gazette*, August 19–26, 1723.

whale was dead. It appears from the court's opinion that the whale sunk or was otherwise lost until two days later when Sturges discovered and brought it to shore with the help of Davis' boat.[11]

Judge Menzies of the Court of Admiralty explicitly recognized the custom of whalemen and the facts in the case at bar in awarding the following shares in the whale: Davis 5/8, Sturges 1/8, and the two boats that assisted Sturges 1/8 each. Each of the awardees were directed to pay a proportionate share of 5 pounds granted to Lambert "towards the Defraying the expense of his cure and as a Moderate gratification and acknowledgmt for his Misfortune." In explaining his decision, Menzies attached particular significance to the fact that Davis was fast to and in control of the whale when the call to Sturges was made and accepted. Menzies also stressed that Davis briefly continued in pursuit of the whale after the accident, returned quickly to the hunt, and that his rope "all the while remained fast to the whale, and that as having a right of Property he did work in bringing the whale to Land." The legal rationale for Menzies' ruling appears to be a combination of custom and common sense. The custom of giving a 1/8 share for assistance is to be honored once it is accepted, even, as in the instant matter, where circumstances necessitated that the first striker detach from the line and seek medical assistance. That Menzies stressed the timely return and further participation of Davis suggests that the customary agreement was not inviolate. It was the obvious and significant manifestations of Davis' continued intention to claim the whale that preserved the agreement. The precise significance of the fact that Davis' line remained fastened to the whale throughout the entire affair is unclear. It appears from Menzies' opinion, however, that once the call for assistance was accepted – as in the Greenland practice of fast-fish, loose-fish when a second boat strikes a fast whale – the continued physical attachment of the first striker was legally irrelevant.[12]

Absent from Menzies' opinion was any reference to the Roman or common law which, as we have seen, figured so prominently in the rulings of British judges in the eighteenth and nineteenth centuries. It may well be that Roman and common law ideas about fastness as evidence of possession constituted the decision's unstated foundation. The language employed by Menzies was certainly consistent with such an understanding of the applicable law. What possibly engaged the court's interest and perhaps led to the litigation was not so much the established principle that honored the rights of a fast boat that calls for assistance, but the status of the two boats that aided Sturges. If the custom of providing a 1/8 share was contractual in nature, how can someone in Sturges' position grant shares in the whale beyond that which he possesses? How can Davis' share, to look at the issue from a different perspective, be

[11] *Davis v. Sturges* in Noble, *A Few Notes on Admiralty Jurisdiction in the Colony and in the Province of the Massachusetts Bay*, 28–30, 29.

[12] *Davis v. Sturges* in Noble, *A Few Notes on Admiralty Jurisdiction in the Colony and in the Province of the Massachusetts Bay*, 28–30, 30.

reduced without his approval? The doctrine of privity of contract would presumably prevent Sturges from binding Davis to an agreement to which he was not a party. Whether the custom of New England whalemen dealt with this situation is unknown. Assuming that it did not, Menzies probably based his decision on what seemed fair. The two boats that aided Sturges were entitled to a reward for their efforts which they believed, in good faith, would garner them each the promised share. While Davis would receive less than he would have under the terms of his agreement with Sturges, considering his need to hastily leave the chase, a 5/8 share of the whale was equitable. This interpretation accords with the thinking of a jurist who saw fit to award Lambert 5 pounds for his medical costs and suffering. Viewed in this manner, Menzies' ruling was similar to those handed down in British cases such as *Hogarth* which used whaling customs as a starting point in crafting a law of whaling.[13]

While the litigants in *Davis v. Sturges* contested a whale taken a short distance from their Barnstable homes, the parties in the 1765 case of *Doane v. Gage* had ventured far from their Cape Cod ports of origin. In the forty-five years that separated these two disputes, much about New England whaling had changed. The shore whaling off the coast of Massachusetts that predominated in 1720 had, by 1765, given way to more distant voyages in search of a quarry that could no longer be found in sufficient quantity so close to home. The increased time and distance necessitated larger ships and the use of whaleboats that could be lowered for the hunt once the desired whaling grounds were reached. Increasingly, after 1720, New England crews headed south to the Carolinas and north to the recently opened Davis Strait fishery. Trips north carried a risk in addition to those attendant to any season whaling. British hostilities with France left New England whalers vulnerable to French attacks they were ill equipped to fend off. The French surrender of Montreal in 1760 effectively reopened Canadian waters to American whalemen for the following season. New Englanders eagerly returned to the Gulf of St. Lawrence and the Strait of Belle Isle situated between Newfoundland and Labrador. In 1761, ten vessels from Massachusetts fished the Gulf of St. Lawrence and Strait of Belle Isle. That figure grew to fifty in 1762 and reached eighty the following year.[14]

Joseph Doane and Lot Gage were captains of two of the one hundred or so whalers that set out from Massachusetts for the 1765 season. Like the overwhelming majority of their colleagues they were bound for the Gulf of St. Lawrence and the Strait of Belle Isle. On June 21, 1765, boats belonging to the ships captained by Doane and Gage were, in company with the crews of other vessels, pursuing the same whale. Asa Nickerson threw a harpoon from one of Doane's boats that either – depending upon the testimony one chooses to believe – missed its target, struck the whale and drew loose before Gage affixed

[13] *Davis v. Sturges* in Noble, *A Few Notes on Admiralty Jurisdiction in the Colony and in the Province of the Massachusetts Bay*, 28–30.

[14] Starbuck, *History of the American Whale Fishery*, 168–73. Dolin, *Leviathan*, 94–101, 124–7.

an iron, or remained in the whale when Gage's harpoon attached. In any event, Nickerson's harpoon was no longer in the whale when Gage killed and took possession of the whale for rendering. While the facts were decidedly ordinary, the legal firepower enlisted to litigate the dispute was not. John Adams, representing Doane, asserted that Nickerson's harpoon was affixed to the whale at the time Gage struck and, thus, according to the customs of the fishery, deemed fast. Notes taken by Adams at the arbitration hearing indicate that he argued that whaling customs "are certain Regulations dictated by observation, Experience, of Common Sense among Whalemen." The customs followed in the Strait of Belle Isle were, Adams continued, "wise, prudent, and equitable." Seeking to convince the arbitrators of the importance of honoring whaling customs, Adams – presaging the argument of Justice Chambre in *Fennings* – explained: "Some Rule, and Law, they must have, to avoid everlasting Contention. What better Rule can they have than this that the first striker shall have the Game."[15]

Whether James Otis and Robert Treat Paine, the prominent attorneys appearing for Gage, agreed with Adams' version of whaling custom is uncertain. In his notes, Adams recorded Paine as asserting that the custom advanced by Doane was odd or, as Adams interpreted that word, unreasonable. To label a custom as deficient in reason was, as the Adams justification for following such a practice indicated, tantamount to dismissing it as lacking the requisite characteristics of an enforceable custom. Paine appeared to favor a different rule that also functioned, more importantly, as Doane's burden of proof. "We are in Possession and Possession is a good Title, untill a Person demands who has an absolute Right." Yet, one of Gage's witnesses provided expert testimony as to the custom of the fishery very similar to that offered by Doane. Custom may not, however, have been central to a resolution of the dispute. The evidence presented by both sides seems to have been directed more at the facts than the law. If Nickerson failed to strike the whale or his harpoon drew prior to Gage's successful thrust, the custom advanced by Adams would not have helped his client. The arbitrators ultimately ruled in favor of Gage. As the rationale for their decision has not survived, one must assume – given the lack

[15] L. Kinvin Wroth and Hiller B. Zobel, eds., *The Legal Papers of John Adams* (Cambridge: The Belknap Press of Harvard University Press, 1965), 2:68–97, 76. In June 1766, Adams brought the matter in state court in Barnstable, but after two years and the taking of numerous depositions, the action was withdrawn. While the reasons for the withdrawal are unclear, Adams, on January 6, 1768 filed the action in the Vice Admiralty Court. The parties subsequently agreed to send the dispute to arbitration. The arbitration hearings apparently lasted six days stretched over a period of several months in the latter half of 1769. The evidence was presented by counsel reading from the depositions and interrogatories of seventy-four witnesses. For a brief discussion of the use of Roman law in this case by John Adams, see Andrew Lewis, "John Adams and the Whale," in *Critical Studies in Ancient Law, Comparitive Law and Legal History; Essays in Honour of Alan Watson*, eds., John W. Cairns and Olivia Robinson (Oxford: Hart Publishing, 2001), 261–6.

of any real controversy over the governing custom – that they simply found Nickerson was not fast to the whale when Gage struck.[16]

The custom of the Gulf of St. Lawrence and Strait of Belle Isle would certainly have been familiar to the parties in *Davis v. Sturges*. Despite the differences between shore-based and pelagic whaling and the passage of forty-five years, Cape Cod whalers still awarded a 1/8 share to boats that answered the call of a fast first striker for assistance. It was, however, in the crucial area of what constituted the fastness of the boat making the call for help that a change in custom can be seen. In *Davis*, the determinative moment was when the call for help was accepted. Davis retained his claim to the whale even though Sturges – having agreed to help – did not strike the whale until after Davis was loose. By 1765, in the waters between Labrador and Newfoundland, the point at which fastness was measured was pushed back to the instant the called second striker affixed his harpoon. This alteration in the custom which favored the boat rendering aid was accompanied by another change that worked very much to the advantage of the first striker. In *Davis*, the first striker came loose when the rope was dropped from the boat. That his harpoon remained in the whale – while mentioned by the court – was not enough to constitute fastness. The custom in *Doane*, testified to by several witnesses and not contested, considered a boat fast so long as the harpoon remained in the whale. In the short-hand account of Adams, Gamaliel and Barzillai Smith explained that there were "Several Instances last year, of Whales struck with only an Iron and naked Warp, and they that struck em again had 1/8, by arbitration." Thus, at some point between 1720 and 1765 Massachusetts whalemen – or at least those fishing in the vicinity of the Gulf of St. Lawrence – had made a transition to the standard of iron holds the whale. This is perhaps the earliest known American application of this norm and demonstrates that it was not, as Ellickson argues, an adaptation made once whalemen – particularly those from Nantucket – began to pursue the faster and more aggressive sperm whale in the emerging Southern fishery.[17]

[16] Wroth and Zobel, *The Legal Papers of John Adams*, 2:82. Evidence that Gage prevailed at arbitration can be found in letters between Joseph Otis and Robert Treat Paine dated December 9, 1769 and January 10, 1770 discussing payment of Paine for his legal services. Joseph Otis was the brother of James Otis and the son of James Otis, Sr. Joseph and his father had a financial interest in Gage's ship. The decision to hire Paine may have been based on his familiarity, as a former whaleman, with the industry or, perhaps, a reflection of the family's concern about the mental health of James Otis, Jr. Stephen T. Riley and Edward W. Hanson, eds., *The Papers of Robert Treat Paine* (Boston: Massachusetts Historical Society, 1992), 2:361, 455. For a 1770 description of Otis' erratic behavior and a general discussion of the treatment of mental illness in this period, see Mary Ann Jimenez, "Madness in Early American History: Insanity in Massachusetts from 1700 to 1830," *Journal of Social History* 20, no. 1 (Autumn, 1986): 25–44, 25.

[17] *Davis v. Sturges* in Noble, *A Few Notes on Admiralty Jurisdiction in the Colony and in the Province of the Massachusetts Bay*, 28–30. Wroth and Zobel, *The Legal Papers of John Adams*, 2:75.

III

An island in the Atlantic Ocean situated about thirty miles south of Cape Cod, Nantucket enjoyed a decided geographical advantage over its competitors in an era of shore-based whaling. Right whales wintering between Cape Cod and the Carolinas preferred to stay within approximately thirty miles of the coast. The necessity of swimming around Cape Cod during migration placed large numbers of right whales on a path within a few miles of Nantucket. Nantucket was also blessed with a ready solution to the perennial labor shortage in the British colonies. Unable to find enough willing men of English descent in the first half of the eighteenth century, the Quaker families that controlled Nantucket whaling turned to the island's sizable Indian population to supply their boats with the muscle required to row after a fleeing whale. The final ingredient for Nantucket's eighteenth-century whaling preeminence was the product of necessity. The sandy soil of Nantucket – the legacy of its formation on the edge of a stalled glacier – was not ideal for agriculture. What little organic material imbued in Nantucket's soil was quickly depleted by farming, the felling of trees, and the needs of a growing populace on slightly less than 50 square miles of land. The early English residents of Nantucket quickly recognized that their long-term survival required that the resources of the sea be exploited. Given all the factors pointing Nantucket toward a future as a whaling power, it is remarkable that it took the English until 1690 – thirty years after their arrival – to successfully recruit a whaleman to teach the techniques of shore-based whaling. As historians have observed ever since, Nantucket quickly took to whaling.[18]

American whaling before the Revolution was mostly conducted off the coast of North America as far south as the Carolinas. Cape Cod's whalemen specialized in following right whales to their feeding grounds up the coast as the summer season progressed. While ships from mainland ports continued throughout the eighteenth century to primarily pursue right whales and, when venturing into the colder waters of the Greenland fishery, bowheads, the whalemen of Nantucket developed an expertise in the hunting of sperm whales. According to the oft told and almost certainly apocryphal tale, Nantucket's

[18] Daniel Vickers, "The First Whalemen of Nantucket," *The William and Mary Quarterly*, Third Series 40, no. 4 (October, 1983): 560–83; Daniel Vickers, "Nantucket Whalemen in the Deep-Sea Fishery: The Changing Anatomy of an Early American Labor Force," *Journal of American History* 72, no. 2 (September, 1985): 277–96; Dolin, *Leviathan*, 63–74; Edouard A. Stackpole, *The Sea-Hunters: The New England Whalemen during Two Centuries, 1635–1835* (Philadelphia: J. B. Lippincott Company, 1953), 17–45. Nantucket's population in the period around 1700 has been estimated at about 300 Englishmen and 800 Indians. Dolin, *Leviathan*, 69; and Vickers, "The First Whalemen of Nantucket," 570, 577. For the standard account of Nantucket's history to which all scholars still turn, see Obed Macy, *The History of Nantucket; Being a Compendious Account of the First Settlement of the Island by the English, Together with the Rise and Progress of the Whale Fishery; and Other Historical Facts Relative to Said Island and its Inhabitants* (Boston: Hilliard, Gray, and Co., 1835).

first successful encounter with a live sperm whale at sea occurred around 1712. Nantucket captain Christopher Hussey was, the story goes, caught in a heavy wind and blown out to sea far beyond the coastal waters plied in the hunt for right whales. Finding themselves in the deep waters frequented by sperm whales, Hussey and his intrepid men took advantage of the opportunity to strike and kill a cachalot. That the small boats employed in shore-based whaling might be blown out to sea is certainly believable. The ability of Hussey and his men to tow a whale to shore the thirty or so miles from waters populated by sperm whales off the New England coast is less plausible. Christopher Hussey's tender age of six in 1712 would seem to deliver to the story – at least in its particulars – a fatal blow. What is certain, however, is that by the second decade of the eighteenth century, Nantucket whalemen were actively pursuing sperm whales.[19]

Hunting for sperm whales in their favored habitats required longer voyages. Given the preference of sperm whales for temperate waters far from chilly New England, Nantucket whalemen quickly adopted – as did their colleagues on the mainland – the European model of ocean going vessels carrying smaller whaleboats and trypots. An industry that began within sight of home expanded dramatically in geographical scope during the eighteenth century. From two to three month trips to Newfoundland in the 1730s to four to five month journeys to the Azores, West Indies, or Davis Strait by the 1760s, American whaling grew to meet an increasing demand at prices that justified the effort. By the end of the eighteenth century, Americans – particularly those sailing out of Nantucket – increasingly embarked on year-long cruises to the coasts of West Africa and Brazil.[20]

American whalemen were always opportunistic. A ship heading north for a season right whale hunting was happy to take a sperm whale should the opportunity arise. Each of the eight Cape Cod ships that ventured to the Strait of Belle Isle in the summer of 1789 returned, for example, with the oil of both right and sperm whales. The same flexibility was also displayed by whalemen whose primary target was the sperm whale. The venerable Nantucket whaleman William Rotch urged the island's Southern fleet in the 1780s to "go after Right Whales if they are more plenty than Sperm." Yet, the fact that New England vessels were willing to take available oil regardless of its source, should not obscure that the destinations preferred by whalemen from particular ports does much to explain why Nantucket became the world's capital of the sperm

[19] Vickers, "The First Whalemen of Nantucket," 562–3; Vickers, "Nantucket Whalemen in the Deep-Sea Fishery," 281–2. For the classic statement of the Christopher Hussey story, see Macy, *The History of Nantucket*, 36. For a recent assessment of the plausibility of the version of events offered by Macy and others, see Dolin, *Leviathan*, 71–2.

[20] Simona L. Perry, Douglas P. DeMaster, and Gregory K. Silber, "The Sperm Whale," *Marine Fisheries Review* 61, issue 1 (1999), 59–74. Vickers, "The First Whalemen of Nantucket," 565–6; Vickers, "Nantucket Whalemen in the Deep-Sea Fishery," 281–2.

whale fishery and a leader in the manufacture of spermaceti candles by the last quarter of the eighteenth century. In 1792, six of the eight ships from the New Bedford area, for which a destination is known, were listed by Alexander Starbuck as headed for the Atlantic. That Starbuck, whose 1876 compilation of American whaling statistics upon which all scholars still rely, considered an Atlantic voyage to be one to the north, is apparent when the destinations he provided for the Nantucket fleet for the same year are considered. Six Nantucket ships went to Walvis Bay off the coast of present day Namibia and another half dozen sailed to the waters of Brazil. A single Nantucket vessel, the *Maria*, rounded Cape Horn and entered the Pacific Ocean. One of the first American whalers to visit the Pacific, the *Maria* returned after more than a year with a substantial cargo of 730 barrels of sperm oil.[21]

On the eve of the Revolution, Nantucket dominated American whaling. For the years 1771 to 1775, Massachusetts ports annually sent out an average of 304 whalers, approximately half of which were based in Nantucket. If only the number of ships bound for the Northern fishery are considered, Nantucket and Dartmouth each provided a third of the total fleet. The Southern fishery presents a very different picture. Nantucket was the home port of 85 of the 121 ships annually heading south. Dartmouth, the second leading point of departure for southbound whalers, could claim an annual average of only 20 ships. Annual averages of oil returned for this period makes Nantucket's dominance of American whaling even more apparent. With 49% of Massachusetts ships accounting for only 35% of the fleet's annual tonnage, Nantucket took 66% of the sperm oil and 52% of the whale oil harvested in an average year between 1771 and 1775. Nantucket's share of all oil brought back to Massachusetts in this period was a remarkable 63% indicating that the island's whalemen were even more productive than they were numerous. When the less detailed figures for all American ports are considered, the importance of Nantucket is only slightly diminished as Massachusetts probably accounted for about 84% of the colonial fleet and around 87% of its entire production of sperm oil.[22]

British whaling in the 1770s and before was an entirely northern enterprise and, as a result, produced only whale oil and bone. The systematic hunting of sperm whales in southern waters was almost exclusively American. Yet, it was the British who constituted the world's primary market for sperm whale products. So long as the American colonies remained British, Nantucket whalemen and London merchants were content with the arrangement. No serious

[21] For the Rotch quote, see Edouard A. Stackpole, *Whales & Destiny: The Rivalry between America, France, and Britain for Control of the Southern Whale Fishery, 1785–1825* (Amherst: University of Massachusetts Press, 1972), 100–1. Starbuck, *History of the American Whale Fishery*, 1:186–9. For the eighteenth-century spermaceti candle industry, see Dolin, *Leviathan*, 109–18; and Stackpole, *The Sea-Hunters*, 28.

[22] Starbuck, *History of the American Whale Fishery*, 1:57.

plans to send British ships to the Southern fishery were in the works before the Revolution. At double the price of whale oil, New Englanders were willing to harvest sperm oil for markets across the Atlantic. With the outbreak of hostilities in 1775, Nantucket's whalemen were in a difficult position. Thirty miles out into the ocean and heavily dependent upon products from England and the mainland which they could not produce, Nantucket's leading whalemen saw few viable options to sending its boats out for oil as they had for generations. In September 1775, Francis Rotch of Nantucket and several merchants from the mainland developed a plan that would allow American whalers to continue in the Southern fishery. They sent out a fleet of whalers to the vicinity of the Falkland Islands to hunt sperm whales. Rather than return to Nantucket at the end of the season, the captains were directed to head for England. Despite the plan to provide oil exclusively to the British, Rotch wanted to establish a base of operations in the Falkland Islands as a plausible show of neutrality. The plan failed when several of the ships were captured by the British navy and one was taken by the American John Paul Jones. While the loyalty of Nantucket to the American cause was called into question when the scheme was revealed, the plan was not a total failure. Francis Rotch, who had gone to England to facilitate the venture, secured the release of his men and ships and thereafter sent some of the Nantucket whalers to the Falkland Islands directly from London. The precise extent or success of Rotch's British operations is unknown, but, in considering the fate of the rest of the American fleet during the war, it could hardly have fared worse. A few Nantucket whalers, having agreed to return its oil to America and managing to secure the required permits and post bond, did set sail for the Southern fishery hoping to evade British ships. Whaling from ports on the mainland, however, ceased almost entirely until near the end of the war.[23]

The disruption in Nantucket's whaling operations during the war provided British merchants who dealt in the products of the Southern fishery an opportunity and a compelling reason to send out their own whaling ships from London. Recognizing that British whaling talent was engaged in the Greenland fishery and military service, London merchants such as Samuel Enderby and Sons turned to the experienced whalemen of Nantucket who – despite the possible consequences of their decision to serve the enemy – were eager to ply their trade. Even though the British were entering a period of dominance in the Northern fishery, they lacked the American expertise in the business of hunting whales in the South. While the actual techniques and skills required for killing whales was remarkably static over the years and throughout the world,

[23] Dolin, *Leviathan*, 119–64; Stackpole, *Whales & Destiny*, 1–13; Dauril Alden, "Yankee Sperm Whalers in Brazilian Waters, and the Decline of the Portuguese Whale Fishery (1773–1801)," *The Americas* 20, no. 3 (January, 1964), 267–88; and Rhys Richards, *Into the South Seas: The Southern Whale Fishery Comes of Age on the Brazil Banks 1765 to 1812* (Paremata: Wellington, New Zealand, 1993).

the dangers and challenges presented by local conditions in the water and on shore were significant. British merchants enjoyed early success in the Southern fishery with captains and crews of Nantucketers and the protection of the Royal Navy. In 1779, John Adams unsuccessfully urged the Continental Congress to send a warship to the Southern fishery to capture some of the London based whalers. With the cessation of hostilities, American whaling began a process of rebuilding its tattered fleet that would not be completed until after the end of the War of 1812.[24]

The steady growth in England's Southern fishery during the 1780s was fueled by the collapse of American competition, a rise in the price of sperm oil, and the passage of favorable legislation. In 1786, Parliament passed an act directed at the "Encouragement of the Southern Whale Fishery." The act, also called the "Fishery Bill," provided premiums for ships returning with the most oil. More importantly, it permitted British whalemen under certain conditions to sail beyond Cape Horn and the Cape of Good Hope and into waters heretofore reserved for the exclusive use of the South Seas Company and the East India Company. This opening of new fisheries in the Pacific and Indian Oceans was deemed crucial to the growth of the industry that had taken notice that whales were becoming less plentiful in waters off the coast of Brazil. From a handful of ships at the beginning of the 1780s, the British Southern fishery fleet had grown by 1793 to sixty-seven ships. While the number of British ships in the Southern fishery remained fairly constant over the next sixteen years – reaching only sixty-nine vessels in 1801 and averaging seventy-two whalers for the period from 1800 to 1809 – the total tonnage of the fleet and the amount of oil taken increased significantly. The rise of British fortunes in the Southern fishery should not obscure the degree to which that success was gained by the labor of Nantucketers. As was pointed out in debate over the Fishery Bill, two-thirds of the men on British whalers bound for the Southern fishery were not subjects of the King. The reliance on foreign expertise is particularly evident when the home ports of the captains and officers commanding British whalers are considered. More than half of all British whalers in the Southern fishery in the early 1790s were captained by men from Nantucket. In the period from 1791 to 1793, as the Southern fishery expanded into the Pacific, a crewmen on an American vessel observed that of the forty ships hunting beyond Cape Horn – twenty-two of which were British – Nantucketers were at the helm of thirty.[25]

[24] Stackpole, *Whales & Destiny*, 1–17, 153; Stackpole, *The Sea-Hunters*, 66–84; and Grandville Allen Mawer, *Ahab's Trade: The Saga of South Seas Whaling* (New York: St. Martin's Press, 1999), 36–52.

[25] Stackpole, *Whales & Destiny*, 112–31, 281–2; Stackpole, *The Sea-Hunters*, 145–63; and Mawer, *Ahab's Trade*, 77–96. Of the other eighteen ships whaling in the Pacific Ocean from 1791 to 1793, seven were out of Nantucket, eight from Dunkirk, and three hailed from other American ports. Dunkirk was established in the whaling trade pursuant to an agreement between the French government and William Rotch after the British rejected the Nantucket whaleman's proposal to establish operations in England. Dolin, *Leviathan*, 171–82.

The speed with which British and American whalers spread throughout the Pacific and Indian Oceans in the 1790s and the first decades of the new century is well illustrated by the opening of the sperm whale fishery in the Galapagos Islands. In January 1793, with a formal declaration of war between the French and the British less than a month in the future, the Enderby whaling firm, in conjunction with the British government, sent a ship to the Pacific to warn whalers of the impending conflict and explore potential whaling grounds. The *Rattler*, captained by an officer from the Royal Navy with extensive experience in the Pacific, reached as far north as Baja California before returning to London in August 1794. News quickly spread in the Anglo-American whaling community that the *Rattler* had found the waters near the Galapagos Islands richly stocked with sperm whales. The fine harbors and the abundance of giant tortoises – viewed by whalemen, in the words of a whaling historian, as "walking larders" – also made the Galapagos Islands a particularly promising new fishery. Hostilities between Britain and Spain generally prevented British whalemen from trading or finding safe harbor in the Spanish ports of Chile and Peru. For British and American ships – the latter tainted perhaps by the close-knit nature of Anglo-American whaling – the Galapagos Islands were a needed refuge after a long trip up the western coast of South America.[26]

A far outpost in the European disputes of the 1790s, the burgeoning Galapagos whalefishery remained, well into the second decade of the nineteenth century, a place where commerce and the specter of armed conflict mixed easily. When the *Caerwent* set sail from London for the Southern fishery in early 1804, Captain Job Anthony carried permission from the East India Company to round Cape Horn and whale as far north as the equator. Anthony also held instructions from the agent for the ship's owner, Lord Camelford, that he should sail in company with the *Cambridge* to the latitude deemed most promising for the taking of whales. Should the two ships belonging to Lord Camelford be separated several rendezvous points, including the Galapagos Islands, were established. Prior to leaving London, the *Caerwent*, like most British ships sailing for foreign ports, was required – pursuant to the Convoy Act of 1798, as amended in 1803 – to give security that it would not depart without the assigned navy escorts providing protection from privateers. Whalers in the Southern whalefishery were not, however, merely targets that required protection. They could turn aggressors in a time of war. Soon after the Spanish declared war on Britain in December 1804, British whalers were issued Letters of Marque and General Reprisal authorizing the taking of ships and

[26] Mawer, *Ahab's Trade*, 86–8; and Stackpole, *Whales & Destiny*, 155–7. For the account of the *Rattler's* captain, see James Colnett, *A Voyage to the South Atlantic and Round Cape Horn into the Pacific Ocean, for the Purpose of Extending the Spermaceti Whale Fisheries, and Other Objects of Commerce, by Ascertaining the Ports, Bays, Harbours, and Anchoring Berths in Certain Islands and Coasts in Those Seas at Which the Ships of the British Merchants Might Be Refitted* (Amsterdam: Nico Israel, 1968), reprint of the 1798 London edition.

goods belonging to the King of Spain or his subjects. On April 13, 1805 a Letter of Marque against Spain was issued for the *Caerwent* and undoubtedly sent to the Pacific; perhaps left in one of the designated locations in the Pacific that served as a post office for Anglo-American whalers.[27]

The apprehension with which the Anglo-American whalemen sailed the waters off Chile and Peru in the first half of 1805 – awaiting word of an official declaration of war with Spain – is evident in the experience of the *Caerwent's* fellow London whaler *Cyrus*. The history of the *Cyrus* and its captain, Nantucketer Paul West, also reveals much about the ease with which national boundaries and affiliations were breached in the whaling industry. The original owner of the *Cyrus* was the Rotch family of Nantucket who operated the vessel as part of its fleet sailing from the French port of Dunkirk. In September 1803, with Paul West serving as first mate, the *Cyrus* was taken by a British frigate off the east coast of Africa. Sold to a London merchant, the *Cyrus* was overhauled and with West in command set sail for the Southern fishery in the summer of 1804. As whaling historian Edouard Stackpole has observed, "The fact that the Nantucketer [West] had sailed under the French flag previously made little difference to either the new owner or the new captain – the *Cyrus* was a whaleship and her master was a whaleman." Judging from the shared intelligence concerning French and Spanish privateers and the movement of whales, the captains of American and other British whalers also found West's personal history of little interest or concern. Of particular interest to West and his peers during the summer of 1805 was whether the much expected war between England and Spain had begun. Finally, in early June, the *Cyrus* encountered six Nantucket whalers bearing the news that war had been declared. Thereafter, the *Cyrus* whaled successfully in the Galapagos Islands while managing to take eight Peruvian fishing boats. The means by which the *Caerwent* learned of the war with Spain or that it had been granted Letters of Marque is unknown. In any event, the *Caerwent* joined the *Cyrus* and other British whalers in the service of the King by pursuing vulnerable Spanish vessels. On September 13, 1805 the *Caerwent* took as a prize the Spanish schooner *Nostra*.[28]

[27] The *Caerwent's* Letters of Marque, license from the East India Company to whale west of Cape Horn, convoy certification, and voyage instructions, Hull Maritime Museum, GB/NNAF/C57178. For the Convoy Act of 1798, see Arthur Herman, *To Rule the Waves: How the British Navy Shaped the Modern World* (New York: HarperCollins, 2004), 402. The *Caerwent* was also issued Letters of Marque against "the French and Batavian Republics, and against the Countries stiling themselves the Ligurian and Italian Republics." As these Letters were issued in 1803 – a year before the *Caerwent* left London – they were likely on board at the start of the voyage. For an example of Pacific whalemen using an overturned barrel to create a protected space for the placing and collection of mail, see Edward J. Larson, *Evolution's Workshop: God and Science on the Galapagos Islands* (New York: Basic Books, 2002), 45.

[28] Stackpole, *Whales & Destiny*, 273–9, 274. Stackpole's earlier volume of whaling history has a slightly different version of the events surrounding how Paul West came to captain the *Cyrus* which includes time in a French prison. Stackpole, *The Sea-Hunters*, 401–3. Information as to

Despite its foray into distant European conflicts and the attendant lure of plunder, the *Caerwent's* primary purpose remained filling its hold with oil. In pursuit of sperm oil selling on the London market at about eighty pounds per ton, the crew of the *Caerwent* some time in the summer of 1805 fell into dispute with the *William Fennings* over the ownership of a sperm whale. The conflict was finally resolved before an English court in the previously discussed matter of *Fennings v. Lord Grenville*. The dispute arose when, as the court set forth the facts, Captain Luce of the *William Fennings* "was engaged in killing a whale." Seeing the opportunity to strike another member of the shoal, Luce affixed a harpoon to a passing sperm whale. Rather than maintain a physical connection to the second target by means of a rope, the *William Fennings* attached to the harpoon a short line with a drogue at the end. The drogue served to mark the movement and tire the whale to which it was tied. Drogues – in various forms and made of different materials – were favored by the native whalers of North America. American whalemen, on occasion, adopted this method of whaling which – for obvious reasons – would be impractical in a fishery that followed a strict interpretation of fast-fish, loose-fish. Luce of the *William Fennings* apparently continued his pursuit of the boat's initial target, letting the second whale tire itself before being chased down and killed. While the statement of facts offered in the case report is very brief, what happened next reveals how the mixture of British and American whalemen in the Southern fishery could lead to confusion, misunderstanding, or even deception. When the second whale was allowed to swim off, "Anthony, the master of the *Caerwent*, who, in consequence of a signal made to him by Luce, followed the fish, and killed it." The signal given by Luce likely evolved from those made by eighteenth-century Massachusetts whalemen calling for assistance. In *Doane*, as we have seen, the price of asking for help was a one-eighth share of the whale. The custom in the Galapagos was that a whaler that killed a whale drogued by another boat was entitled to half of the proceeds from the catch.[29]

Although the Galapagos custom of sharing whales has struck whaling historians as anomalous, it appears to be an extension of earlier American practices. Given the prevalence of American masters in the Southern fishery, the continuation of the American custom of sharing whales is logical. What is not clear is the degree to which in the Galapagos a call or signal for assistance was made before a second whaler took up the hunt. The published report implies that Anthony's pursuit of the whale was the direct result of Luce's signal. Even if that aspect of earlier Massachusetts custom had not been carried south, the

the *Caerwent's* capture of the Spanish schooner is contained in orders issued on October 5, 1805 by Captain Anthony warning of the punishments that would be meted out if crew members get drunk or misbehave while plundering enemy ships. Hull Maritime Museum, GB/NNAF/C57178.

[29] *Fennings v. Lord Grenville*, 1 Taunt. 241 (1807). The nationalities of Luce and Anthony are unknown.

signal might have remained more as an acknowledgment of the second striker than a condition precedent to taking a share in the whale. That the court believed that the Galapagos custom dated to 1792 – at the opening of that fishery – is a clear indication that the practice of sharing drogued whales was common in the Southern fishery well before Anglo-American whaling moved up the coast to the equatorial waters plied by Captains Luce and Anthony.[30]

Whether Luce's signal was necessary for the granting of a half share or – as is more likely – the vestige of an earlier practice, Luce certainly expected Anthony to honor the custom and split the whale. Anthony's response to the signal is unknown, but – whatever it was – it did not apparently give Luce concern that a dispute was in the offing. As a signatory of an agreement with other captains in the Galapagos Islands that a boat that kills a whale need not give half of the proceeds to the whaler that affixed a drogue, Anthony very likely knew that Luce had not bound himself to such an arrangement. Captain Anthony might have reasoned that a feigned acceptance of Luce's signal would allow him to take the whale and that the master of the *William Fennings* would not press for his share of the whale. It is also possible that Anthony believed, as counsel for the *Caerwent* argued unsuccessfully at trial, that a new custom that eschewed the splitting of whales had gained general assent and should be enforced even if Luce had not indicated his agreement in writing.[31]

Although the agreement signed by Captain Anthony has not survived, a similar document signed by seven captains whaling in the Galapagos Islands during the 1801 season is extant. Of the ten articles, several deal directly with how disputes over the ownership of whales are to be resolved. The provision of the 1805 agreement governing drogued whales was likely similar to the terms of the 1801 contract which provided as follows: "If a boat Strikes a Whale & parts & then Strikes another, the first whale so Struck becomes the property of Any Boats who shall kill her; but must Return the Craft. Yet, if the first boat is following his whale & hath not been fast to another he Claims the whale, even though she is killed by Another Ship's boat." In both the 1801 and 1805 agreements the mere affixing of a harpoon – with or without a drogue – did not reserve the first striker's rights. What the 1801 contract contained and the lost 1805 agreement likely included was the proviso that if a boat has a harpoon in a whale and maintains pursuit, it is entitled to the fish so long as it is not fast to another whale. The purpose of this emerging custom in the Galapagos Islands and probably the entire Southern fishery was clearly to prevent a single ship from making claims to multiple whales at the same time. While a drogued whale was easier to capture, it must still be pursued and killed. Honoring the claims of a whaleman who was not in active pursuit resulted in escaped whales or, worse yet, a decomposed animal without any value.[32]

[30] *Fennings v. Lord Grenville*, 1 Taunt. 241.
[31] *Fennings v. Lord Grenville*, 1 Taunt. 241.
[32] Stackpole, *Whales & Destiny*, 279–80, 388–9.

Robert Ellickson has suggested that the Galapagos Islands custom of splitting sperm whales between the first striker and the boat that ultimately killed the whale was an adaptation made to improve the efficiency of whalers pursuing a speedy and dangerous foe that travels in large groups. He believes that the even split encouraged whalemen, on one hand, to strike as many whales as possible and, on the other hand, to kill all harpooned or drogued whales. While a single whaler would not get the entire whale, it would get a large enough share to justify the effort of striking or killing their prey. The custom, in Ellickson's estimation, resulted in a sensible division of labor, was simple to administer, and helped "foster a spirit of cooperation." The importance Ellickson places on the type of whale hunted in the creation and evolution of customs is, however, misplaced. Whaling techniques were remarkably consistent throughout the seventeenth, eighteenth, and nineteenth centuries. Whalemen threw harpoons which were attached to ropes at whales. Ideally, the harpoon remained fast to the whale and the rope served to keep the boat in close enough proximity to make the kill. Ropes broke, harpoons drew loose, and circumstances frequently required that lines be cut. Such was the lot of whalemen in all fisheries. Sperm whales are certainly fast and, when struck, ornery. While often characterized as docile, bowheads and right whales are clearly not passive when attacked. The ice in the northern fisheries frequented by bowheads often severed lines or caused harpoons to break free. Whalemen in icy waters often cut ropes lest they be dragged by sounding bowheads beneath sheets of ice to a likely death. The change from fast-fish, loose-fish to iron holds the whale was not made because whalemen approached the challenges of bowhead and sperm whaling differently. British whalemen in the Greenland fishery were slowly moving in the first half of the nineteenth century – with, as we have seen, judicial assistance – toward iron holds the whale while pursuing bowheads. New Englanders by the end of the eighteenth century were, as evidenced in *Doane v. Gage*, honoring the first striker whose harpoon remained affixed to an otherwise loose right whale. Perhaps, most telling is that when British and American whalemen hunted the Southern fishery, they took both sperm and right whales without any alteration in their technique. American whalemen simply carried their preference for iron holds the whale to the Southern fishery in the 1790s.[33]

Attempting to trace the development of whaling custom in the Southern fishery is complicated by its vast expanse, mixture of British and American whaleman, the presence of European settlers hunting mostly in bays, and the existence of colonial courts. In the Greenland fishery – at least after 1800 – British and Americans predominated and, for the most part, hunted amongst their countrymen. The Americans, for example, were never active in the Spitzbergen area and were minor players in the Davis Strait. The legal rights of native hunters were never considered. When Anglo-American whalers,

[33] Ellickson, *Order without Law*, 191–206, 202.

however, took to the Southern fishery in growing numbers in the last quarter of the eighteenth century it became increasingly difficult to define with any confidence which ships were American and which were British. To which nation does a ship of British registry owned by London merchants, but captained and largely manned by Americans belong? While the owners would be legally correct in thinking of their ship as British, an American crew would – at least in the early years of the Southern fishery – likely conduct themselves at sea according to the customs established in places like the Gulf of St. Lawrence. Add to this, such complications as American owners of British registered ships and Nantucketers operating out of the French port of Dunkirk and the international character of whaling at the turn of the nineteenth century becomes apparent. Four of the seven captains who agreed as to how they would conduct themselves in the Galapagos Islands for the 1801 season were from Nantucket. Three of the ships sailed from London, two from Nantucket, and two from the Welsh port of Milford Haven which had been established as a whaling community by Nantucketers by way of Dartmouth, Nova Scotia.[34]

As the pelagic whaling customs in the Southern fishery developed in the first years of the nineteenth century the general American preference for iron holds the whale came to predominate in places like the Galapagos Islands where Nantucketers dominated the industry. Despite the differences between what the British court in *Fennings* recognized as the Southern custom of sharing drogued whales and the 1805 agreement which limited a boat's right to claim multiple whales, both rules were, in fact, variations of iron holds the whale. The primary distinction between the two practices set forth in *Fennings* concerned whether or not the first striker must remain in active pursuit of a marked whale in order to retain a possessory interest in the catch. While iron holds the whale was generally followed in the pelagic branch of the Southern fishery, the rules in the bays fished primarily by European colonists were more varied. Each bay in New Zealand, for example, had, by the early 1840s, developed its own rules concerning the capture of whales. Edward Jerningham Wakefield, the generally dissolute son and nephew of major proponents of British colonization and sovereignty over New Zealand, explained that the practice in most of the island's whaling bays honored a first striker whose distinctly marked harpoon remained affixed to its target. The bay whalers of Australia also generally adhered to the principles of iron holds the whale. Tasmania, however, adopted the practice of fast-fish, loose-fish. In 1838, Tasmanian legislation provided that a whale is fast and belongs to the first striker so long as the harpoon remains in the whale and the attached line continues in the control of the boat's crew. "But," the statute continued, "where the harpoon so struck shall break or become disengaged from the fish, or the instrument shall remain fast in the fish but the line affixed thereto shall break or be wholly run out, or the fish from any

[34] For the establishment of whaling in Milford Haven, see Stackpole, *Whales & Destiny*, 197–226.

other causes not be in the power or under the management of the striker such fish shall be considered a loose-fish and become the property of the actual taker." In the British colony of Cape of Good Hope – yet another bay whaling outpost of the Southern fishery – the local court declared in a whale capture case: "There is no proof of any rules, laws, or usage regulating the whale fishery in Simon's Bay or in the sea adjoining the coasts of this Colony."[35]

Pressing into the bowhead and right whale grounds of the North Pacific at mid-century, American whalemen continued the custom of iron holds the whale. They did so, not, as Ellickson argues, as a holdover from sperm whaling that was poorly adapted to the hunting of bowheads, but as continuation of a custom that was reasonably well suited for any species of whale chased by men in boats with ropes, harpoons, and lances. The relative paucity of foreign competition assured that American practices would dominate Pacific whaling. With the collapse of British whaling and the absence of European colonial whalemen in the Pacific the old standard of fast-fish, loose-fish ceased to be a practiced custom and became – as evidenced by Melville's usage – a sort of generic term to describe how American whalemen resolved disputes at sea over contested whales.[36]

[35] Edward Jerningham Wakefield, *Adventure in New Zealand from 1839 to 1844* (Christchurch, New Zealand: Whitcombe and Tombs Limited, 1908), 229. For the Wakefield family and their involvement in New Zealand, see E. Irving Carlyle, "Wakefield, William Hayward (1803-1848)," in Sidney Lee, ed., *Dictionary of National Biography* (New York: The Macmillan Company, 1899), 456–7; and E. Irving Carlyle, "Wakefield, Edward Gibbon (1796-1862)," in Sidney Lee, ed., *Dictionary of National Biography* (New York: The Macmillan Company, 1899), 449–52. For the Western Australia practice of iron holds the whale at King George's Sound (present day Albany), see *The Perth Gazette and Western Australia Journal*, May 4, 1833. For the Tasmanian law of 1838, see Kathryn Evans, *Shore-Based Whaling in Tasmania: Historical Research Project* (Hobart, Tasmania: National Parks and Wildlife Service, 1993), I:50. The 1838 act amended an 1835 law which provided that when a harpoon remained in the whale, but the line broke or came free, the prey would be split between the first striker and the boat that dealt the fatal blow. The 1838 act was repealed in 1842 as Tasmanian shore and bay whaling was in its last days. Mawer, *Ahab's Trade*, 101–4. The reason for the amendment is unknown. Whatever the reason, it was likely directed at further protecting the interests of Tasmanian shore and bay whalers from the encroachment of opportunistic pelagic whalers who often hunted in bays when in the vicinity at certain times of the year or when their usual prey proved particularly elusive. For complaints made by colonial shore and bay whalemen about American and French pelagic whalers imposing on their grounds, see Kylli Firth, "'Bound for South Australia:' 19th Century Van Diemen's Land Whaling Ships and Entrepreneurs," Flinders University Maritime Archaeology Monographs Series, no. 9, 2006, 3–4, accessed August 30, 2008, www.ehlt.flinders.edu.au/archaeology/department/publications/MAMARS/MAMS%20PDF/9%20Firth%202006_final.pdf. Background on the legislative process for Tasmania and the other Australian colonies is provided at Michael Quinlan, "Making Labour Laws Fit for the Colonies: The Introduction of Laws Regulating Whalers in Three Australian Colonies 1835–1855," *Labour History* 62 (May 1992): 19–37. *Langley v. Miller* (Supreme Court of the Cape of Good Hope, 1848), in James Buchanan, ed., *Cases Decided in the Supreme Court of the Cape of Good Hope, as Reported by the Late Hon. William Menzies, Esquire* (Cape Town: J. C. Juta, 1870), 584–9.

[36] Ellickson, *Order without Law*, 191–206.

5

Laws of honor

As Herman Melville and William Scoresby recognized, the whaling customs of fast-fish, loose-fish, and iron holds the whale were never strictly applied at sea. A fairness standard which Melville called justice and Scoresby deemed the "laws of honor" was also recognized. Whalemen were never able or never cared to explain exactly what they meant by fairness. Captains serving as expert witnesses in American courts often testified as to their personal sense of what was fair with the caveat that they spoke only for themselves. Yet, these inchoate ideas about just behavior were shared by whalemen to the degree that they were recognized when followed or breached. Major breaches such as failure to render assistance in times of danger or the refusal to provide accurate intelligence as to conditions or the location of whales were serious matters because they touched on the safety of crews and the economic imperative of killing and rendering whales.

Notions of fairness are always difficult to define and, perhaps, even more difficult to apply in the courtroom. While American courts were content to base a decision on a whaling custom that was – like fast-fish, loose-fish, or iron holds the whale – reducible to a relatively coherent rule, vague statements of fairness were quickly dismissed. The obligation to provide accurate – or at least not misleading – information to competitors which served as a sort of Golden Rule for whalemen was, for example, viewed by courts as unenforceable. Where whalemen saw foundational principles, judges observed behavior that, while perhaps admirable, was neither contractually binding nor, in the breach, tortious. This inability of courts to grasp the importance of fairness in resolving property disputes in combination with an abiding preference for settled rules does much to explain why legal professionals and institutions either failed or chose not to recognize the fluidity of customs and the improvisational nature of how whalemen went about their business.

I

The focus of the American whaling industry shifted to the bowhead and right whale fisheries of the North Pacific in the 1840s. Prior to 1840, there were no American whalers in the North Pacific. By 1855, half of all American whale ships could be found in latitudes above 50 degrees north. As with much of whaling history, the push toward Arctic waters at mid-century can be explained by a combination of supply and demand. Sperm whales in their Pacific habitat south of 40 degrees north grew increasingly difficult to catch after 1820. In the 1820s, a ship could expect to approach its capacity of sperm oil. By the late 1840s, the number of barrels of sperm oil per shipping ton had dropped in the previous twenty-five years from approximately 7½ to less than 5. The increasing scarcity of sperm whales coincided with a spike in the value of whale bone. The dollars per pound real value of unprocessed whale bone for the New Bedford market rose in the five year period beginning in 1839 from 0.16 to 0.52. A shift in women's fashion to skirts made fuller by whale bone and a drop in sperm oil demand after 1840 drove whalemen north to the baleen whale laden waters of the North Pacific. Fisheries off the coast of Kamchatka and in the Gulf of Alaska flourished by 1845, the year the first American whaler ventured into the Sea of Okhotsk. The expansion of the North Pacific fisheries was complete with the 1848 passage of the *Superior* of Sag Harbor through the Bering Strait and into the Arctic Ocean.[1]

Whalemen intensively hunted bowheads in the Sea of Okhotsk for about twenty years. By 1871, commercial whaling in the Sea of Okhotsk was finished. No American ships slipped through the Kuril Islands that summer to chase bowheads around the Shantar Islands or any of the bays favored by hunters. The sudden rise and precipitous fall of the Sea of Okhotsk fishery is, in many respects, the quintessential whaling story. Whalemen, always anxious to find the next harbor or patch of ocean teeming with cetaceans, rushed to the Sea of Okhotsk, fished with tremendous vigor, and then moved on when other grounds showed greater promise.

In the North Pacific, bowheads can be found in latitudes from 54 to 75 degrees north. The Sea of Okhotsk, which extends from the southern tip of the Kuril Islands at a latitude of 44 degrees north to the far reaches of the Gulf of Shelekhova at about 63 degrees north, is – despite its latitude – subject to ice cover of up to 85% in late February and March. By comparison, ice cover in the Western Arctic's Bering Sea is limited mostly to shelf areas. Strong, persistent

[1] John R. Bockstoce, *Whales, Ice, and Men: The History of Whaling in the Western Arctic* (Seattle: University of Washington Press, 1986), 21–9; George W. Schuster, "Productivity and the Decline of American Sperm Whaling," *Environmental Affairs* 2 (1972–1973): 350–1; Eric Hilt, "Investment and Diversification in the American Whaling Industry," *The Journal of Economic History* 67, issue 2 (June 2007): 292–314, 295; and Lance E. Davis, Robert E. Gallman, and Karin Gleiter, *In Pursuit of Leviathan: Technology, Institutions, Productivity, and Profits in American Whaling, 1816–1906* (Chicago: University of Chicago Press, 1997), 69, 376.

winter winds over the Siberian land mass and a high percentage of fresh water of lower salinity account for the extensive annual ice cover in the Sea of Okhotsk. Open water can be found in the winter on the eastern edge of the Sea of Okhotsk near the Kuril Islands and the coast of Kamchatka. The northern portion of the sea is also subject to wind driven patches of open water. Each year a hole in the ice can be found in the vicinity 55 degrees north and 145 degrees west, created when a warm current meets the slope of the Kashevarov Bank.[2]

While their winter home in the Sea of Okhotsk is unknown, bowheads can be found each spring in the vicinity of the Kashevarov Bank. Like all polynyas – areas of open water surrounded by ice that appear annually in the same location – Kashevarov Bank in winter teems with life across multiple trophic levels. Bowheads find here the necessary open water for respiration and, in spring, a bounty of zooplankton. Nineteenth-century whalemen were well aware that their quarry gathered early in the hunting season near Jonas Island, the landmark by which they referred to the waters of the Kashevarov Bank. In 1863, Captain Otto Lindholm explained that bowheads gravitated to Jonas Island each spring in pursuit of the zooplankton that he believed was delivered annually to the region by the prevailing currents in the Sea of Okhotsk.[3]

A typical whaling season in the Sea of Okhotsk began with the departure of the fleet from Hawaii in March or early April. By the end of May, some of the fleet were chasing whales near Jonas Island and some headed into higher latitudes bound for Gizhignskaya Bay or what whalemen called the Northeast Gulf. The main body of the fleet generally made its way by June 20th

[2] Sue E. Moore and Randall R. Reeves, "Distribution and Movement," in John J. Burns, J. Jerome Montague, and Cleveland J. Cowles, eds., *The Bowhead Whale* (Lawrence, KS: The Society for Marine Mammalogy, 1993), 313–86; Takeshi Okunishi, Michio J. Kishi, Akihiro Shiomoto, Hitoshi Tanaka, and Toshihiko Yamashita, "An Ecosystem Modeling Study of Spatio-Temporal Variations of Phytoplankton Distribution in the Sea of Okhotsk," *Continental Shelf Research* 25 (2005): 1605–28; James D. Hays and Joseph J. Morley, "The Sea of Okhotsk: A Window on the Ice Age Ocean," *Deep-Sea Research I* 50 (2004): 593–618; and Victor V. Lapko and Vladimir I. Radchenko, "Sea of Okhotsk," *Marine Pollution Bulletin* 41, nos. 1–6 (2000): 179–87.

[3] For an understanding of polynya formation and their biological richness, see Francesca Vidussi, Suzanne Roy, Connie Lovejoy, Marie Gammelgaard, Helge Abildhauge Thomsen, Beatrice Booth, Jean-Eric Tremblay, and Behzad Mostajir, "Spatial and Temporal Variability of the Phytoplankton Community Structure in the North Water Polynya, Investigated Using Pigment Biomarkers," *Canadian Journal of Fisheries and Aquatic Sciences* 61, issue 11 (November 2004): 2038–52; Oliver F. Berreville, Alain F. Vezina, Keith R. Thompson, and Bert Klein, "Exploratory Data Analysis of the Interactions Among Physics, Food Web Structure, and Function in Two Arctic Polynyas," *Canadian Journal of Fisheries and Aquatic Sciences* 65, issue 6 (June 2008): 1036–46; and Andrey Y. Shcherbina, Lynne D. Talley, and Daniel L. Rudnick, "Direct Observations of North Pacific Ventilation: Brine Rejection in the Sea of Okhotsk," *Science* 302, issue 5652 (December 12, 2002). Otto Lindholm, "Whales and How Tides and Currents in the Sea of Okhotsk Affect Them," January 2, 1965, Scripps Institution of Oceanography Technical Report, accessed August 19, 2007, www.repositories.cdlib.org/sio/techreport/21.

through the melting ice from the vicinity of Jonas Island the five or so degrees of longitude west to the port at Ayan. The ice was not sufficiently degraded at this point in the season to permit passage of ships south along the coast into Udskaya Gulf. Whale boats, manned by crews of six, were, instead, dispatched to chase the bowheads as they made their annual migration toward the Shantar Islands. The small boats allowed whalemen to navigate or pull south, as needed, through and across the remains of the winter ice cover. Bowheads killed at this time could be anchored in the frigid water or hauled ashore to await a reunion with the ships made possible by rising temperatures over the course of the next few weeks. The Okhotsk fleet would then settle in for a season of whaling lasting until the middle or so of October.[4]

Bowheads migrated to the vicinity of the Shantar Islands each year in pursuit of food. Supporting the nineteenth-century observations of Lindholm and other whalemen, scientists are, at present, revealing the mechanisms in the Sea of Okhotsk which in Spring drive the ice and zooplankton in tandem south along the Siberian coast. Gravitational circulation, tides, river inflow, and ice melt all play a part in concentrating zooplankton in the shallow coastal water south and west of the Shantar Islands. Whalemen moved in and out of the many bays and around the numerous small islands in this compact portion of the Sea of Okhotsk throughout the summer in search of their prey. By the middle of October, weather conditions – not the departure of the whales – drove whalemen out of the Sea of Okhotsk and back to Hawaii or another fishery in a more temperate region.[5]

[4] For a description of when and where whales were hunted in the Sea of Okhotsk, see, in addition to Lindholm, "Whales and How Tides and Currents in the Sea of Okhotsk Affect Them," Kim E. W. Shelden and David J. Rugh, "The Bowhead Whale, Balaena Mysticetus: Its Historic and Current Status," *Marine Fisheries Review* 57, issue 3/4 (1995): 1–20; Moore and Reeves, "Distribution and Movement," 357–60; C. M. Scammon and Edward D. Cope, "On the Cetaceans of the Western Coast of North America, *Proceedings of the Academy of Natural Sciences of Philadelphia*, Vol. 21 (1869): 13–63; and Charles M. Scammon, *The Marine Mammals of the North-Western Coast of North America* (New York: Dover Publications, Inc, 1968), 52–65.

[5] Lindholm, "Whales and How Tides and Currents in the Sea of Okhotsk Affect Them"; Scammon and Cope, "On the Cetaceans of the Western Coast of North America"; Shelden and Rugh, "The Bowhead Whale"; and Moore and Reeves, "Distribution and Movement," 357–60. The oceanographic literature on the Sea of Okhotsk is large and obviously quite technical. Particularly useful for understanding the migration of bowheads from Jonas Island to the Shantar Islands, are Jun Ono, Kay I. Ohshima, Genta Mizuta, Yasushi Fukamachi, and Masaaki Wakatsuchi, "Amplification of Diurnal Tides over Kashevarov Bank in the Sea of Okhotsk and Its Impact on Water Mixing and Sea Ice," *Deep Sea Research Part I: Oceanographic Research Papers* 53, issue 3 (March 2006): 409–24; Genta Mizuta, Kay I. Ohshima, Yasushi Fukamachi, Motoyo Itoh, and Masaaki Wakatsuchi, "Winter Mixed Layer and Its Yearly Variability Under Sea Ice in the Southwestern Part of the Sea of Okhotsk," *Continental Shelf Research* 24, issue 6 (April 2004): 643–57; and, in particular, Konstantin A. Rogachev, Eddy C. Carmack, and Michael G. G. Foreman, "Bowhead Whales Feed on Plankton Concentrated by Estuarine and Tidal Currents in Academy Bay, Sea of Okhotsk," *Continental Shelf Research* 28, issue 14 (August 2008): 1811–26.

II

The first American whaling dispute to go to litigation in the nineteenth century was *Taber v. Jenny*. The long journey to the United States District Court in Boston began with a rather common occurrence in the Sea of Okhotsk. A boat crew, having killed a bowhead, anchored their prize for recovery when weather conditions improved and their ship could be reached. When another ship found the whale and discovered markings on the harpoon identifying the slayer, fairness clearly required that the bowhead be returned to its rightful owner which was anchored nearby. The decision of the intervening vessel to keep the bowhead did not, however, incite the opprobrium of other captains. The slayer, having earlier lied to the intervener as to the presence of whales and their hunting success, forfeited the moral high ground and – in the opinion of the community of captains – ceded its right to the whale.

Fog was general and dense along the southern coastline of the Sea of Okhotsk from the Shantar Islands to Cape Elizabeth at 5:00 AM on the morning of July 23, 1852. John Leonard, the twenty-four year old New Bedford man serving as first mate of the *Hillman*, had spent the previous evening on shore with men from his ship and the *Massachusetts*. Through the fog, five or six whales were observed within a quarter mile of the shore. The men quickly pushed off in four boats to pursue their prey. Almost immediately, the boat commanded by Leonard struck a bowhead whale. For the next two-and-a-half to three hours, the whale – with the attached boat in tow – moved away from shore. Leonard recounted in a deposition taken two years later that they finally managed to kill the whale in fifteen fathoms of water about six miles out into the Sea of Okhotsk. The chances of quickly finding either the *Hillman* or the *Massachusetts* in the thick fog were not favorable. Rather than drag the dead whale to shore or search for their ships, Leonard decided to anchor the carcass where it had been taken. A seventy to eighty pound anchor was affixed to the whale by a doubled seventy-five fathom length of rope. A piece of pig iron was thereafter attached to the line to determine whether the whale was drifting in the current. Satisfied after nearly an hour of observation that the whale was securely anchored, the boats of the *Hillman* and the *Massachusetts* returned to shore to await improved weather. Before making their departure, a waif made of blue cotton sewn to a cedar pole was affixed to the dead whale to aid in its later recovery.[6]

[6] Depositions of John W. Leonard, George D. Whitney, John D. Maxfield, and James Fraser. Case file, Taber v. Jenney, United States District Court for the District of Massachusetts, Special Court, December 1855, United States National Archives, Northeast Region, Waltham, Massachusetts (National Archives, Northeast Region). The caption used by the District Court provided the correct spelling of defendant Levi Jenney's name. When the court's opinion was published, Jenney's name was misspelled. *Taber v. Jenny*, 23 F. Cas. 605 (US Dist. Ct., D. MA, 1856). As the case has, as a result of this error, come to be known as *Taber v. Jenny*, that spelling is followed when referring to the court's published decision.

Whalers in the bays of the Sea of Okhotsk frequented by bowheads often sent boats out to search for their quarry. Instead of waiting to sight a whale before lowering their boats as was common in fisheries far from land, Okhotsk whalemen – like their colleagues plying the bays and harbors of Australia, New Zealand, and the Galapagos Islands – found that they could cover more ground and kill more whales if multiple crews were set free to roam in protected waters. In the Sea of Okhotsk the small whaleboats had the additional advantage of being able to maneuver through the ice and reach the open water near the shore where bowheads could be found early in the hunting season. The breakup of the ice in the southern Sea of Okhotsk's Shantar Bay was far enough advanced by the end of June to permit the entry of the season's first whaleboats. Whaling ships were forced to wait sometimes weeks before conditions permitted a reunion with their boats. With the melting of the ice and clearer weather as the season progressed, ships were free to enter Shantar Bay or proceed to the more northerly bowhead grounds of the Sea of Okhotsk. Although whaling from the ship was practiced more frequently later in the season, boats were still sent out with provisions sufficient for extended periods of hunting. Boats, like those of the *Hillman* and the *Massachusetts*, often camped on the beach during these forays from their ships.[7]

The fog remained thick in the southern Sea of Okhotsk on the morning after Leonard and his men had anchored their whale. Captain Avery F. Parker aboard the *Zone* estimated in his journal that – judging from the sound of waves breaking on the beach – his ship was about a mile from land. Only the tops of nearby mountains were visible through the fog. At 8:00 AM that Saturday morning, Parker observed two boats heading toward shore. During a brief discussion, Parker learned that one of the boats was from the *Massachusetts* and that the other, under Leonard's command, belonged to the *Hillman*. As was common in all whale fisheries, information as to the location of whales was exchanged. Leonard told Parker that the single whale they had spied since entering the bay was observed heading out to sea. This, of course, was not true. In addition to the whale he had killed the previous day, Leonard was almost certainly aware that other whales had been taken by boats belonging to the *Hillman* and the *Massachusetts*. Leonard stated in his deposition that on the morning of Saturday the 24th he had returned to the *Hillman* and observed a whale being towed to the ship. After eating breakfast aboard ship, Leonard headed back to the location near the shore where he had struck the bowhead the day before. It was at this point that Leonard presumably encountered Captain Parker. Why then did Leonard decide – at least about his

[7] For contemporary descriptions of whaling in the Sea of Okhotsk, see "Northern Whaling," *Overland Monthly and Out West Magazine* 6, no. 6 (June, 1871), 548–54; and Edward Dusseault, "Recollections of Other Days," *Ballou's Monthly Magazine* 49, no. 6 (June, 1879): 556–61. Dusseault recounts an 1857 whaling trip to the Sea of Okhotsk.

own success – to lie to Parker? Leonard likely wanted to discourage Parker from remaining in what seemed to be a promising portion of Shantar Bay.[8]

Leonard did, however, provide Parker with a useful bit of accurate information. The *Massachusetts* and the *Hillman*, Leonard revealed, were anchored about eight miles off shore. Hoping perhaps that the *Massachusetts* and the *Hillman* carried mail from home, Parker immediately set the *Zone's* course in the direction of the anchored ships. The *Zone* came across the *Massachusetts* at about 10:30 AM and Parker discovered that Captain Bennett did, indeed, have mail for him. Aboard the *Massachusetts*, with the *Hillman* in sight, Bennett – repeating Leonard's lie – offered that they had neither struck nor killed any whales. Bennett added that the whales that they had seen were wild which, in the parlance of whalemen, indicated that boats were having trouble getting within striking distance. It is possible that Bennett, who was well aware of the whales taken by the boats of the *Hillman*, was simply speaking of the *Massachusetts*, but this seems unlikely as he also revealed that the two vessels had entered into an agreement of mateship. Mateship was a common arrangement where two ships decide to share their catch at the end of a predetermined period. Parker, if not already aware of this pact, would likely have surmised as much from the close cooperation he had observed between the boats of the *Hillman* and the *Massachusetts*. The visit aboard the *Massachusetts* was brief and by 11:00 AM the boat which had taken Parker to see Bennett was making its way back to the *Zone*.[9]

The fog was so thick that Parker's crew was uncertain as to the *Zone's* exact location. While attempting to find its way back to the *Zone*, the boat came upon a dead bowhead. Pulling closer, the crew observed the waif pole affixed by Leonard. Boatsteerer James Fraser testified that Parker, given his conversations with Leonard and Captain Bennett, appeared surprised to come upon a whale that – bearing no signs of putrification – had been recently killed. Captain Parker removed the pole and directed that the boat return to the *Zone* which was now visible through the shifting fog at a distance of about a half-mile. Parker boarded the *Zone* and ordered first mate Ephraim Gifford to return to the whale and tow it back to the ship. Once the whale was relocated with some difficulty, Gifford and the boat's crew discovered that a line was already attached. The men began hauling in the line. Their first pulls met with little resistance, but the work soon became more burdensome. An anchor of the

[8] Journal of Avery Parker, July 24, 1852. KWM Log # 211, New Bedford Whaling Museum, New Bedford, Massachusetts (NBWM). Depositions of John Leonard and Owen W. Crosby, Case File, Taber v. Jenney, National Archives, Northeast Region.

[9] Journal of Avery Parker, July 24, 1852, NBWM. Depositions of John Smith, George D. Whitney, and James Fraser, Case File, Taber v. Jenney, National Archives, Northeast Region. For a contemporary legal description of mateship, see David Roberts, *A Treatise of Admiralty and Prize: Together with Some Suggestions for the Guide and Government of United States Naval Commanders in Maritime Wars* (New York: Hurd and Houghton, 1869), 417.

sort commonly used to secure dead whales for later recovery was soon brought to the surface, cut loose, and placed in the boat. Gifford and some of the men, recognizing that whoever had killed the whale would likely return once conditions improved, hesitated before Gifford finally directed that they bring the prize back to the *Zone* as ordered. The sense that they were perhaps acting in a less than honorable fashion in taking a whale rightfully belonging to a competitor was likely heightened by the fact that although Gifford's men could not see the *Hillman*, its crew could be heard at work. With the whale alongside the *Zone*, the process of slicing the blubber into pieces suitable for boiling – known as cutting in – began at around 1:30 PM. When the whale was cut into, two irons bearing the markings of the *Hillman* were discovered. Parker faced a decision. The thickness of the fog had allowed the *Zone* to recover the whale without drawing the attention of either the *Hillman* or the *Massachusetts*. Despite the proximity of the *Hillman*, Parker decided not to inform its captain, Christopher Cook, of the whale's discovery. By 6:30 PM, the crew had finished cutting in and within two hours the tryworks were lit and the lengthy process of boiling the whale's blubber had begun.[10]

Leonard, in the meantime, had not given up the search for his missing whale. After coming upon the *Zone* in the spot just off shore where he struck the whale, Leonard, in company with other boats, followed the path he had taken the day before in pursuit of the bowhead. By evening, Leonard was in the vicinity of where he believed he had anchored the whale. The fog, having begun to lift, allowed a view of the *Zone* boiling its prize. Disappointed, Leonard returned to the *Hillman* to report his failure to find the whale and the news that one of their competitors was boiling. Whether Leonard had any suspicion that the whale he had killed was presently being reduced on the *Zone* to barrels of oil and stacks of bone is unknown. That Leonard harbored such thoughts is likely as the following afternoon he decided to pay a visit to the *Zone*. Aboard the *Zone*, Leonard asked about the whale which was still being boiled. Captain Parker, perhaps wishing to indicate that the whale had drifted far from its original anchorage, stated that they had found it about twenty miles out into the Sea of Okhotsk. While estimates offered by the crew of the *Zone* as to the exact location of their discovery varied from ten to twenty miles from shore, it is difficult to believe that Parker was being honest with Leonard. Parker, to be sure, did not know his precise position. His journal contains much speculation as to his coordinates for the days surrounding these events. He did, nevertheless, imply in his journal that he encountered the *Massachusetts* – as Leonard said he would – at a distance of about eight miles from shore. Even in Parker's remarkably disingenuous pleadings before the United States District Court, he averred the distance to be ten to fifteen miles. There can be no doubt, however, that both Parker and Gifford were lying when they

[10] Journal of Avery Parker, July 24, 1852, NBWM. Depositions of John Smith, Owen W. Crosby, George D. Whitney, and James Fraser, Case File, Taber v. Jenney, National Archives, Northeast Region.

told Leonard that no anchor was recovered with the whale. Perhaps as a ploy to show his good faith, Parker ordered that the irons found in the whale be brought on deck for Leonard's examination. Leonard immediately recognized the markings on the iron to be those of both the *Hillman* and his particular boat. Whatever Leonard reported upon his return to the *Hillman* sent Captain Cook – in company with Captain Bennett – to confront Parker about ownership of the now disputed whale.[11]

When Cook and Bennett reached the *Zone* around 11:00 PM on the 25th the tryworks were still in full operation boiling out the last of the bowhead's blubber. Bennett left little doubt that this was not to be a friendly social call – or gam in the whalemen's argot – when he came aboard the *Zone* declaring, in the words of a witness, that "he wanted that whale and have it he must." Parker provided no satisfaction and by several accounts the exchange was heated. Bennett and Cook, Parker noted in his journal, spoke to him "in a very ungentlemanlike manner," accusing him of stealing the whale. "Finally my grey hairs and age," Parker wrote, "was all that protected me."[12] The *Zone's* second officer, George Whitney, corroborated Parker's description of the encounter, overhearing Bennett say "if it was not for the grey hairs in your head, there would be some blood spilt." Bennett blustered that with two anchors affixed to the whale it could not have drifted. Parker, in response, called Bennett's bluff. What, Parker inquired, had become of the second anchor? Whitney, testified that this query appeared to deflate Bennett's rage. Bennett's reaction was that of someone caught in a lie. He undoubtedly knew that Leonard had attached a single anchor. Bennett must have suspected – correctly – that Parker was also lying when he stated that he did not have the *Hillman's* anchor. If this portion of the argument had the *pro forma* quality of an exchange that had been repeated in some form or another by generations of whalemen, Parker struck at the heart of the dispute and the central paradox of whaling custom when he asked Bennett why he had lied in failing to acknowledge that the *Hillman* and the *Massachusetts* had struck and killed at least one whale. Whalemen – as Scoresby had recognized forty years before and a half a world away – are supposed to behave like Christian gentlemen, while, at the same time, beating their competitors. The deceptions of Leonard and Bennett clearly violated the first half of Scoresby's whaling ethic in service of achieving the second. Bennett replied – in a lame defense exposing his violation of custom – that "he did not know as he had any business to let Capt. Parker or any body else know whether he had got any whales, or any thing of the kind."[13]

[11] Journal of Avery Parker, July 24–25, 1852, NBWM; Depositions of John Leonard; and Avery F. Parker, Answer to the Libel and Complaint of Henry Taber et al., Case File, Taber v. Jenney, National Archives, Northeast Region.

[12] Journal of Avery Parker, July 25, 1852, NBWM; and Deposition of George D. Whitney, Case File, Taber v. Jenney, National Archives, Northeast Region.

[13] Deposition of George D. Whitney, Case File, Taber v. Jenney, National Archives, Northeast Region.

Yet, Parker had also violated the expectation of honorable behavior in not notifying the nearby *Hillman* as soon as the whale was discovered. Parker justified his decision to keep the whale on two grounds. He observed that the whale had drifted many miles from where it was killed and that the *Hillman* would never have found it in the fog. Given that Parker had discovered the whale approximately fifteen minutes after leaving the *Massachusetts*, such a claim was, at best, disingenuous. Parker's second justification was based on whaling norms. He wrote in his journal that "we have him nearly all boiled out and I don't consider they have any claim on him, according to established rules." A longstanding tradition dating back at least to the early nineteenth-century Greenland fishery provided that a ship finding a drifting dead whale was entitled to as much oil as it could boil out before the whaler that inflicted the mortal wound appeared and made a verifiable claim. Verification was generally made by examining the markings whalemen made on harpoons in contemplation of such situations. The key aspect of this custom was that the whale must be drifting. A drifting whale was one that would escape utilization if not recovered before the blubber turned rancid. The single rule of whaling universally recognized in all fisheries in every period was, of course, that not a drop of oil or a pound of bone should go to waste. A securely anchored whale would, on the other hand, likely be retrieved by the boat that affixed the anchor. While determining exact coordinates was difficult, experienced mates were generally successful in returning to a location where a whale was anchored. Even if Parker believed that the bowhead had been drifting, he was aware of which ship had killed it once the *Hillman's* harpoons were brought on deck.[14]

Zone boatsteerer John Smith testified that after Bennett and Cook left the ship without satisfaction, Parker decided to remain off shore to avoid contact with the other whalers anchored near the coast. The clear implication of Smith's testimony was that Parker was concerned that the community of whalemen in the Sea of Okhotsk would disapprove of his actions. While Parker may have harbored some concerns for his reputation, he did not behave like a man in fear of retaliation or the contempt of his fellow captains. On the following Saturday, Parker took a boat ashore and procured a "mess of fish" from the *Trident* and the *Cicero*. Sunset the following day was marked with tea aboard the *Zone* in the company of Captains Blackmer, Taber, and Churchill. When the *Zone* lost its anchor a few days later while Parker and most of the crew were out in boats pursuing a whale, the *Erie* and the *Trident* answered a call of distress and rendered assistance. Parker returned the favor when, shortly

[14] Journal of Avery Parker, July 25, 1852, NBWM. Deposition of Abraham W. Pierce, Case File, Swift v. Gifford, United States District Court, March 1872, National Archives, Northeast Region; Deposition of George Whitney, Case File, *Taber v. Jenney*, National Archives, Northeast Region.

thereafter, the *Zone* alerted the *Trident* to the location of one of its whales that had drifted far from its original anchorage.[15]

It is, of course, possible that Parker was trying to salvage his reputation by a conspicuous show of collegiality. More likely, however, is that Parker was confident that a significant number of his fellow captains would – if placed in the same situation – have also taken the whale. Bennett's deception violated the fundamental duty of every captain to supply accurate information concerning the location of whales to his fellow mariners. Presented with the opportunity to exact a measure of revenge and fill his hold with oil and bone at Bennett's expense, Parker recognized that his own violation of whaling customs would be understood, if not applauded, by his peers. Parker did not have to guess how his actions would be received. The constant interaction between captains in the Sea of Okhotsk and in Hawaii allowed Parker to immediately gauge where he stood in the community. John R. Holt, captain of the *Hunter*, explained that the dispute was "common talk" whenever captains met that summer in the Sea of Okhotsk. Discussion of the incident continued when the fleet returned to Hawaii in Autumn and as many as forty or fifty captains gathered and, in Holt's words, "loafed together." While the tenor of those discussions is unknown, Holt assured Parker, both at sea and in port, that he would have acted in the same manner. Holt even told Captain Bennett that he believed Parker to have been in the right in taking the whale.[16]

Parker's confidence in the reaction of the whaling community did not extend, however, to how this matter would be viewed by a court in Boston should the litigation threatened by Cook and Bennett be pursued. Parker told Smith and other crewmen that he was going to understate the amount and keep separate some of oil taken from the disputed whale. Thus, Parker explained to his crew, in the event of an adverse result in court, he would have some oil for his trouble.[17]

III

When attempts to resolve disputes at sea or in winter ports such as Honolulu failed, the next step was often negotiations between owners. If these efforts proved fruitless and one party was not willing to concede in the belief that the aggravation of the conflict was greater than the value of the disputed whale, arbitration before a panel of whaling captains was the last step before filing a court action. As *Taber v. Jenny* demonstrates in the breach, owners generally accepted the decisions of the arbitrators. The initial acquiescence of Taber and

[15] Journal of Avery Parker, July 31, August 1, August 3, August 6, August 7, and August 8. 1852, NBWM; Deposition of Owen W. Crosby, Case File, Taber v. Jenney, National Archives, Northeast Region; and Deposition of John Smith, Case File, *Taber v. Jenney*, National Archives, Northeast Region.

[16] Deposition of John R. Holt, Case File, *Taber v. Jenney*, National Archives, Northeast Region.

[17] Deposition of John Smith, Case File, *Taber v. Jenney*, National Archives, Northeast Region.

Company – the *Hillman's* owners – to a negative arbitration ruling faded as two pieces of information throwing the fairness of the proceedings into question was brought to their attention. While Taber and Company was willing to accept a ruling which they considered incorrect, they were not disposed to accede to one reached through deception.

On March 17, 1854 the *Hillman* returned to New Bedford after an absence of nearly three years. Captain Cook either immediately informed the ship's agents and owners – Henry Taber, John Hunt, and William Taber of the firm Henry Taber and Company – about his dispute with the *Zone* or had done so previously by letter. The owners of the *Zone* were contacted and a parol or oral agreement was quickly reached to settle the dispute by arbitration. Pursuant to the agreement, each side would choose a man familiar with the whaling trade to act as a referee. Taber and Company selected George S. Tooker. As captain of the *Martha*, Tooker had recently returned from a whaling trip to the North Pacific. Levi Jenney, agent and part owner of the *Zone*, selected – at Captain Parker's suggestion – Captain Holt. From their conversations in the Pacific and back in New Bedford, Parker was certainly aware that Holt would be a friendly arbitrator. After expressing reluctance and gaining assurances that the decision of the referees would be final, Holt agreed to serve. On March 28th, Tooker and Holt heard evidence in the New Bedford office of Taber and Company. Captains Cook and Parker provided the bulk of the testimony. A few additional questions were directed at *Hillman* second mate George Nye and one of the two *Zone* greenhands with the last name of Ritchie.[18]

Having heard the evidence, Tooker and Holt retired to render a decision before a freshly kindled fire. It was soon apparent that the two referees viewed the matter differently. Captain Tooker credited Cook's testimony that the whale had been anchored and believed that the *Hillman* was entitled to a share of the bowhead bearing its harpoons. Holt, claiming that in his experience a

[18] Depositions of John R. Holt, Levi Jenney, Jr., Joseph Tripp, and Isaiah Terry, Answer to the Libel and Complaint of Henry Taber et al., Case File, *Taber v. Jenney*, National Archives, Northeast Region. For details on the 1851–4 voyage of the *Hillman*, see Alexander Starbuck, *History of the American Whale Fishery From its Earliest Inception to the Year 1876* (New York: Argosy-Antiquarian Ltd., 1964), 2:478–9. For Captain Tooker and the *Martha*, see Starbuck, *History of the American Whale Fishery*, 2:480–1. For Captain Holt and the *Hunter*, see Starbuck, *History of the American Whale Fishery*, 2:478–9. A greenhand was the name given to the least experienced members of a whaling crew. There were two greenhands on the 1851–4 voyage of the *Zone* with the last name of Ritchie: F.H. Ritchie and Leon H. Ritchie. The New Bedford Library, New Bedford, Massachusetts maintains a searcheable online database of crew lists at www.newbedford-ma.gov/library/whaling-archives/. For an understanding of the history of arbitration in the resolution of commercial disputes, see Eben Moglen, "Commercial Arbitration in the Eighteenth Century: Searching for the Transformation of American Law," *Yale Law Journal* 93 (November 1983): 135–52; Henry Horwitz and James Oldham, "John Locke, Lord Mansfield, and Arbitration During the Eighteenth Century," *The Historical Journal* 36, no. 1 (March, 1993): 137–59; and Bruce H. Mann, "The Formalization of Informal Law: Arbitration Before the American Revolution," *New York University Law Review* 59 (June 1984): 443–81.

single anchor was not sufficient to hold a whale in the Sea of Okhotsk, argued that the *Zone* should be awarded the entire whale. The precise basis for Holt's view of the matter is not entirely clear. In asserting that he had lost whales in the Sea of Okhotsk secured with two anchors, Holt either thought that the whale was – as Parker testified – adrift or that the currents in those waters precluded a reasonable dependence that such a catch could be found hours later. The latter belief does not, by necessity, require that the whale had actually drifted. Holt seemed to suggest that the likelihood of such a whale being dragged far from its original anchorage was sufficient to deprive the slayer of his prize. In asserting that the whale should be given to the *Zone*, Holt affirmed the industry's interest in maximizing the catch. Had Parker deferred to the *Hillman's* superior claim and allowed the whale to remain in the water, the poor weather conditions and the nasty currents in the Sea of Okhotsk might well have allowed the carcass to be carried beyond the reach of any hunter.[19]

Failing to agree upon the proper disposition of the whale, Tooker and Holt – in accordance with the terms of the arbitration agreement – went in search of a third person to break the tie. Holt suggested that Captain Seth Blackmer would, if available, be an appropriate choice given his 1852 experience in the Sea of Okhotsk. If Holt was, as Taber and Company later suspected, chosen by Parker because his favorable view of the dispute was known, Blackmer may also have been selected for the same reason. Within days of the whale's taking, Blackmer and two other captains enjoyed tea on the *Zone* with Parker. The dispute was certainly discussed at this meeting and Blackmer likely expressed his opinion. Blackmer hesitated to serve, but relented when told that no other master was readily available and that the parties awaited a decision. The three captains proceeded to the house where Holt boarded to discuss the evidence and arguments. Tooker assumed the role of Captain Cook in presenting the *Hillman's* case and Holt took the part of Parker in explaining the arbitration hearing to Blackmer. Captain Blackmer concurred with Holt in believing that the whale should be awarded to the *Zone*. Tooker – having never sailed the Sea of Okhotsk – deferred to the experience of Blackmer and Holt in those waters, permitting the panel to render a unanimous decision in favor of Captain Parker. Holt and Tooker returned to the office of Taber and Company to present the verdict. Jenney asked Taber if he was satisfied with the decision and the latter replied – to the best of Holt's recollection – that he was.[20]

[19] Deposition of John R. Holt, Case File, *Taber v. Jenney*, National Archives, Northeast Region.

[20] Deposition of John R. Holt, Case File, *Taber v. Jenney*, National Archives, Northeast Region. Journal of Avery Parker, August 1, 1852, NBWM. Seth Blackmer had a long career as a whaling captain. For his 1850 to 1853 trip to the North Pacific as captain of the *Erie*, see Starbuck, *History of the American Whale Fishery*, 2:470–1 and the New Bedford Library, New Bedford, Massachusetts online database at www.newbedford-ma.gov/library/whaling-archives/. George Tooker would, unfortunately, later experience firsthand the difficulties of whaling in the Sea of Okhotsk. In 1864, he was captain of the *Mercury* when it was lost in the North East Harbor of the Sea of Okhotsk. Starbuck, *History of the American Whale Fishery*, 2:592–3.

The written decision of the panel was brief. The first of two sentences directed the owners of the *Zone*, without explanation, to provide the *Hillman* with a replacement boat anchor, line, and craft. Reaching the gravamen of the dispute, Captains Holt, Blackmer, and Tooker explained that the whale "was legally taken as Capt. Bennet (sic) gave Capt. Parker to understand that he had no whale at all." The issue of the fastness of a whale anchored in the Sea of Okhotsk which animated the initial deliberations of Tooker and Holt was not mentioned. It seems unlikely, given his earlier insistence on the difficulty of anchoring a whale in those waters with even two anchors, that Holt suddenly abandoned this line of argument. Perhaps, Holt – in seeking to find a rationale upon which all three captains could agree – shifted his grounds to an argument that was based more on the general ethics of whaling than the peculiarities of currents in the Sea of Okhotsk. Having never fished in the Sea of Okhotsk, Captain Tooker might have felt more comfortable with punishing the *Hillman* for Captain Bennett's violation of the rule that a whaleman should provide accurate intelligence to his competitors. Parker must have felt a measure of vindication. His decision to take the whale had been ratified by a trio of his fellow captains.[21]

If, as Holt testified, Taber expressed his satisfaction with the arbitration proceedings, what caused him and his partners to file a complaint five months later in United States District Court? The limited nature of the testimony and the prospect of a quick and inexpensive resolution does much to explain the appeal of arbitration. While there is no way to know how many cases were heard before panels of captains, the evidence from litigated matters suggests that it was fairly common and almost always resolved the dispute. Why then, to put the matter in different terms, did the issue suddenly strike Taber and Company as worthy of the cost, time, and effort of litigation? The likely answer begins with the unbidden visit by *Zone* boatsteerer John Smith to the New Bedford office of Taber and Company in mid-August 1854. Smith revealed that he had been aboard the *Zone's* boat that had discovered the *Hillman's* whale and that the bowhead had, indeed, been securely anchored. As the disputed whale had not – according to Smith – been adrift, any appeal by the *Zone* to the custom cited by Parker governing whales floating freely would be legally irrelevant. The principals of Taber and Company likely reasoned that with Smith's testimony their chances of success in court warranted the expense of bringing suit. In addition, Taber learned within a week of the arbitration that Captain Holt was friendly with Parker and had – previous to the hearing – opined that the whale belonged to the *Zone*.[22]

21 Deposition of John R. Holt, Case File, *Taber v. Jenney*, National Archives, Northeast Region. A copy of the arbitration decision is attached to Holt's deposition.

22 Depositions of John R. Holt and Gideon Carnell and Libel and Complaint of Henry Taber, John Hunt, and William G. Taber v. Henry Taber and Company, Case File, *Taber v. Jenney*, National Archives, Northeast Region.

There was, up to this point, no indication that lawyers had been consulted. Clearly, there was no attorney involvement in the arbitration hearing. Once Taber and Company engaged, however, the services of the law firm of Eliot and Pitman, the matter began its path – as disputes often do when lawyers get involved – to the Boston courtroom of District Judge Peleg Sprague. Part of the calculation of Eliot and Pitman in filing this action within a week or so of the boatsteerer's revelation would have certainly included that Smith – having left the *Zone* mid-voyage on less than congenial terms – would be subject to a brisk cross examination. Smith's deposition, taken two days after the complaint was sworn and a week before the deponent's return to sea, suggests that Parker had disparaged his boatsteerer's whaling skills. Smith denied that he held any grudge against Parker and that, in fact, he had "always liked the man first rate since I have been with him." Despite his protestation that he spoke with the owners of the *Hillman* only out of a sense of duty, Smith's actions were probably not driven by a desire that whaling customs be vindicated. Prior to visiting Taber and Company, Smith called upon the *Zone's* agent and part-owner, Levi Jenney. Smith testified that he simply wished to inform Jenney of his decision to reveal to Taber the circumstances surrounding the taking of the disputed bowhead. While there is no evidence that Smith asked Jenney for money or anything else in return for silence, Smith did state on cross examination that they discussed Parker's poor opinion of his former boatsteerer and Jenney's offer that forbearance would be rewarded with a desired position on a future voyage. Smith, in any event, did admit recently telling an acquaintance that he would make the *Zone* give up the whale taken two years earlier in the Sea of Okhotsk.[23]

The slow progress with which this dispute made its way to a judicial resolution does much to explain why so few whaling contests were litigated in nineteenth-century American courts. As whaling voyages grew longer in the nineteenth century, it was increasingly difficult – if not impossible – to assemble a full complement of witnesses at any one location or time. When whaling was a part-time occupation in the eighteenth and very early nineteenth centuries, crew members could be rounded up with relative ease when the fleet returned in company at the end of the season. In the Greenland fishery of the Scoresbys, for example, the creeping formation of ice forced all ships to make the relatively short trip back to northern England and Scotland each August. Even if the British ships and crews of the Greenland fishery were fitted out for commercial service during the winter, the costs and uncertainties of proceeding to trial were minor compared those facing Taber and Company in 1854. Having drawn up the complaint, the attorneys representing the *Hillman* acted quickly to take the deposition of Smith, their star witness, before shipping out on a voyage that might last years. Owen Crosby was deposed four days later on the eve his

[23] Starbuck, *History of the American Whale Fishery*, 2:478–9. Deposition of John Smith and Libel and Complaint of Henry Taber, John Hunt, and William G. Taber, Case File, *Taber v. Jenney*, National Archives, Northeast Region.

departure. Other witnesses testified as they became available or were deemed necessary over the next two years. Recognizing that most witnesses would not be available to testify before a jury, depositions were taken with the intention that they would be read into the trial record. Defense of this action was doubtless complicated by its inception prior to the *Zone's* May 5, 1855 return to Massachusetts. Fortunately, Captain Parker had returned to New Bedford in 1853 after falling ill at sea and was available to assist in the pleadings and defense strategy.[24]

IV

Once lawyers got involved in the dispute and the action shifted to the United States District Court in Boston, the language and the entire tenor of the matter changed. Whalemen no longer controlled the proceedings and having ceded power to the legal system, the men who made their living from the oil and bone of whales were subjected to what must have seemed a rather esoteric discussion of Roman law and a curious application of the law of salvage. Whaling custom and practice – the law operating in the Sea of Okhotsk and in the office of Taber and Company – was dismissed by the court after a perfunctory discussion. It is likely that the litigants were surprised by the way in which the lawyers and Judge Sprague transformed the meaning of their dispute. No dispute over the possession of a whale had, after all, been litigated in a Massachusetts courtroom in the lifetime of any of the participants. For whalemen, the disagreement was about who should reap the benefits of the catch. The court was, on the other hand, more interested in the point at which a possessory interest in property ripens into an inviolate form of ownership. Judge Sprague was little concerned with the fate of a single bowhead carcass or even hundreds of anchored or drifting whales. It was more important to the court that property rights be upheld than that the whaling industry be allowed to operate in a manner that advanced the interests of participants.

Confirmation that the language in District Court would be very different from that expressed at sea and during arbitration was provided by the legal argument advanced by Eliot and Pitman on behalf of Taber and Company. Rather than point to the customs of the industry, counsel for Taber and Company turned to the same Roman law discussions of how one comes to own wild animals cited in the earlier British cases. In directing Judge Sprague to Book II, Title I, sections 13 to 16 of *The Institutes of Justinian*, Eliot and Pitman asserted that by killing the contested bowhead the crew of the *Hillman* had exercised to the greatest degree possible the control needed to acquire possession of the whale. While the parties likely found the Latin phrases such as *res nullius* that were surely employed by those before the bar alien, the concepts were, of course,

[24] Depositions of John Smith, Owen W. Crosby, and John R. Holt, Case File, *Taber v. Jenney*, National Archives, Northeast Region; and Starbuck, *History of the American Whale Fishery*, 2:484–5.

familiar. Fast-fish, loose-fish was, after all, a more picturesque way of identifying the point at which control over a whale comes to signify possession which will ultimately ripen into ownership. Counsel agreed that in killing the whale, the *Hillman* acquired ownership. The issue was whether the *Hillman* relinquished that ownership when it left the whale anchored, but unattended.[25]

Lincoln Brigham, representing the *Zone*, argued that the accepted custom of whaling was that a ship was entitled to any whale discovered adrift so long as the carcass was cut into before the slayer appeared and made a supportable claim. Judge Sprague determined that although the custom was successfully proven, it was inapplicable to the facts in the case. The whale, Sprague believed, was not adrift when discovered by Captain Parker and his crew. Judge Sprague allowed that, at most, the whale might have dragged the anchor a short distance. Even had the currents driven the whale and anchor a considerable distance from their original moorings, the result would be the same. There was, Sprague explained, no evidence of any usage covering a situation where an anchored whale was carried a great distance. Listening to Judge Sprague's ruling in court or reading his opinion, Captain Parker must have determined that the jurist had not understood what whalemen meant when they spoke of a drifting whale. For Sprague, a whale was adrift when it was not fastened to an anchor or some other object employed for the same purpose. An anchored whale could be pushed hundreds of miles out to sea and still not, in Sprague's estimation, be adrift. Sprague's insistence on this particular definition of a drifting whale was to any whalemen – and particularly one experienced in the notoriously strong currents of the Sea of Okhotsk – ridiculous. As referee turned trial witness John Holt testified, even two anchors could not guarantee that a whale in the Sea of Okhotsk would be found where it was originally secured. Whalemen cared little what a carcass dragged. A whale in motion was in danger of escaping the trypot. The whole point of the custom concerning rights to a drifting whale was that it should be utilized. The slayer's right to his prize lasted only so long as he remained able to quickly render his catch. In the Boston courtroom of Peleg Sprague the continued attachment of an anchor signified the preservation of the slayer's property rights. In the Sea of Okhotsk, the need to turn a profit and the race against decomposition rendered Sprague's expansive notion of property rights in an anchored carcass antithetical to the business of whaling.[26]

[25] The opinion of Judge Sprague does not indicate the argument advanced by counsel on behalf of Taber and Company. For the argument of Pitman and Eliot, see *Henry Taber et al v. Levi Jenny, et als*, *The Monthly Law Reporter* (May, 1856), 27–36, 28. For the translation of Justinian used by Pitman and Eliot, see Thomas Collett Sandars, *The Institutes of Justinian; with English Introduction, Translation, and Notes* (London: John W. Parker and Son, 1853), 181–2.

[26] *Taber v. Jenny*, 23 F. Cas. 605. Deposition of John R. Holt, Case File, *Taber v. Jenney*, National Archives, Northeast Region. Sitting on the bench of the United States District Court in Boston, Judge Peleg Sprague presided over many maritime cases. He was credited by one contemporary as having "substantially created the law applicable to whalemen, whalers, and the whaling business." Roberts, *A Treatise on Admiralty and Prize*, 419. Prior to his 1841 appointment to

When Judge Sprague pondered the facts and law in the case at bar, his inclination was to look for certainty and universally accepted legal principles. Marking a slain whale with a waif and harpoons bearing the ship's initials and taking great care in securing an anchor, constituted – in Sprague's estimation – "unequivocal marks of appropriation." The whale, at this point, became "the absolute property of the *Hillman*." For Captain Parker and the other whalemen plying the waters of the Sea of Okhotsk, absolute ownership in a whale did not vest until it was cut into and its blubber boiling. Whalemen were painfully aware that, prior to that point, there was much that could separate a ship from its putative catch. It was this uncertainty in bringing a slain whale aboard ship for its bubbling denouement and the omnipresent desire to maximize the industry's profit that conditioned whalemen to see disputes over ownership in less absolute terms than those announced from the bench of a federal court. In the fog of the Sea of Okhotsk – both literal and metaphorical – Avery Parker could, in good faith, reason that custom concerning drift whales and the lies of Leonard and Captain Bennett were enough to justify taking a bowhead which he, in other circumstances, might have returned to the *Hillman*.[27]

The basis for Judge Sprague's certainty that the *Hillman* had acquired "absolute and unencumbered" ownership of the disputed bowhead was his application of the law of salvage. Salvage is the branch of maritime law that establishes the rights of one who saves a ship or its goods at sea or on the coast from loss or destruction. It is an axiomatic tenet of salvage law that the salvor gains only a lien on the rescued property in return for his service. Ownership remains in the original holder of the property. Sprague reasoned that a slain whale marked with an anchor, waif, and harpoons was no different than a boat abandoned after encountering some difficulty at sea. Things lost at sea remain the property of the original owner even if all hope of recovery has vanished and recovery efforts have been abandoned. While the amount of compensation for a salvor's efforts depends, in large measure, on a subjective measurement of such factors as risk encountered and the owner's attempt at recovery, the fundamental relationship between the parties in a salvage case and the subject property is constant: ownership never trades hands.[28]

the United States District Court in Boston, Sprague studied at the Litchfield Law School and served Maine in the House of Representatives and the Senate. He retired from the bench in 1865. For details of Sprague's life, see his obituary in *The Harvard Register* 2, no. 5 (November 1880): 232.

[27] *Taber v. Jenny*, 23 F. Cas. 605.

[28] For contemporary discussions of the law of salvage, see Francis B. Dixon, *Abridgement of the Maritime Law; Comprising General and Particular Average, Adjustment, Abandonment, Bottomry, Collision, and Salvage to Which is Added, the General Duties of Masters and Owners, with a Copious Appendix Consisting of Several Useful and Legally Approved Forms* (New York: Charles T. Evans, c. 1857), 110–22; and William Marvin, *A Treatise on the Law of Wreck and Salvage* (Boston: Little, Brown and Company, 1858), 105–80.

Judge Sprague's application of salvage law to the proceeds from a dead whale was not without precedent. In 1831, the British High Court of Admiralty affirmed the ancient principle that a whale that washed ashore or was caught in coastal waters was – as a royal fish – the property of the crown. All other parties involved in such a whale's discovery or capture were entitled to claims of salvage. Far from the British coast and royal prerogative, the application of salvage law to whaling was less certain. The *Swan*, a Hull whaler trapped in the Davis Strait in October 1837, was freed the following May by the cutting of a channel through the ice of nearly a quarter-of-a-mile by the crews of three other ships. When one of the rescuing ships claimed salvage, the *Swan* argued that it was the custom of whalemen to provide "gratuitous assistance" whenever colleagues were discovered in a dangerous situation. The High Court of Admiralty, without deciding whether such a practice existed, determined that the facts in the case at bar could not support application of the alleged custom. Judicial enforcement of the custom and the resulting abrogation of salvage law required, in the court's estimation, that the *Swan* and its rescuers be engaged in a common enterprise and in a position to render aid to one another. The *Swan's* near fatal entrapment in the ice of the Davis Strait left it unable to provide the sort of reciprocal assistance required to place ships outside the law of salvage by virtue of a special arrangement or understanding born of a shared endeavor.[29]

The High Court of Admiralty revisited the issue of whaling custom and salvage a few years later. In *The Harriot*, Judge Stephen Lushington determined that South Seas whalemen customarily eschewed claims of salvage when rendering aid to one another. On November 2, 1838 the *Harriot* – a British whaler – ran afoul of a coral reef as it entered the port of Honolulu. The crew of a second whaler, the *Folkestone*, answered the call for assistance and helped lead the *Harriot* to a safe anchorage. The *Harriot* rejected the *Folkestone's* request for compensation as a salvor on the basis of two customs. It was the custom of the port of Honolulu for the harbormaster to dispatch ships to assist, without compensation, other vessels that encountered difficulty in making their entrance. While this custom was local and applied to all ships, the *Harriot* also invoked the practice universally observed by whalemen of rendering aid to their whaling colleagues without remuneration. Affidavits were produced on behalf of the *Harriot* from merchants, ship owners, and whaling masters that the international community of whalemen had followed this custom for many years. None of these witnesses could, in fact, recall a whaler claiming salvage

[29] *The Lord Warden and Admiral of the Cinque Ports v. H.M. in his Office of Admiralty (In the Matter of a Whale)*, 166 Eng. Rep. 304, 2 Hagg. 438 (1831). *The Swan*, 166 Eng. Rep. 499, 499, 1 W. Rob. 68, 69 (1839). For a general introduction to the British High Court of Admiralty in this period, see F. L. Wiswall, Jr., *The Development of Admiralty Jurisdiction and Practice Since 1800: An English Study with American Comparisons* (Cambridge: Cambridge University Press, 1970).

for services provided to a whale ship in distress. Having avoided a ruling in *The Swan* on the existence and legality of a whaling custom renouncing claims of salvage, Judge Lushington faced the issue directly in ruling that the *Folkestone* was not entitled to compensation as a salvor. "It appears to me," Lushington observed after noting the *Folkestone's* failure to cite a case of salvage between whalers, "to be scarcely possible, considering the immense number of vessels engaged in this trade, that some such demand should not have been preferred, unless it were generally understood that it could not be successfully maintained; or, in other words, that a custom to the contrary was in existence."[30]

Whether Judge Sprague was aware of the British cases concerning salvage rights between whalers is unknown. His contemporary on the federal bench, Judge William Marvin of the Southern District of Florida, cited *The Swan* and discussed the issue in an 1858 treatise on the law of wreck and salvage. Judge Marvin explained that in the slave trade and other dangerous pursuits, ships "do not stand so independently of each other as vessels meeting accidentally, and it may therefore be fairly supposed that they feel the policy and duty of rendering material assistance, and a smaller compensation is sufficient." In particular trades, Marvin indicated, this idea of a shared pursuit in perilous circumstances has led to customs where no remuneration is sought when one vessel comes to the aid of another. In addition to South Sea whaling, Marvin cited the steamboats on the Mississippi River that eschew salvage for rescuing a colleague that has run aground. Sprague, an experienced admiralty court jurist, was likely aware of this line of salvage cases. His decision not to discuss the whaling exception to salvage law was probably twofold. Most importantly, the *Zone* did not claim salvage. Judges, as a general proposition, are loathe to raise claims and possible defenses not advanced by counsel. In addition, the *Zone* had not acted as a salvor. Captain Parker's intent in taking the whale was clearly to appropriate it for his own use, not save it for the *Hillman*. Sprague's discussion of salvage law was not, in any event, directed at explaining the relationship between the disputing ships. Salvage law – for Sprague – provided a way of thinking about property rights in goods lost at sea. That whalemen might have decided among themselves not to claim salvage, was of no concern to Sprague. Absent a clear custom to the contrary – which Sprague found not to exist in the case of a whale remaining affixed to an anchor – general legal principles of property ownership must prevail.[31]

[30] *The Harriot*, 166 Eng. Rep. 636, 639, 1 W. Rob. 440, 447 (1842). For a clearer understanding of the facts and procedural history of *The Harriot*, see *Times* (London), January 30, 1841; *Caledonian Mercury* (Edinburgh), May 28, 1842; and *Times* (London), June 4, 1842. For the long career of Admiralty Judge Stephen Lushington, see J. H. Baker, "Famous English Canon Lawyers: IX Stephen Lushington, D. C. L. († 1873)," *Ecclesiastical Law Journal* 4, issue 19 (July 1996): 556–65; and S. W. Waddams, *Law, Politics and the Church of England: The Career of Stephen Lushington, 1782–1873* (New York: Cambridge University Press, 1992).

[31] Marvin, *A Treatise on the Law of Wreck and Salvage*, 126–7.

The *Zone's* attempt to turn the court's attention to Parker's reliance on the false information provided by Captain Bennett was summarily dismissed by Judge Sprague. Sprague observed that as Captain Cook of the *Hillman* was not present during the exchange between Bennett and Parker, his legal right to the disputed whale could not be impaired. The agreement of mateship between Cook and Bennett was not sufficient, in Sprague's analysis, to convey legal rights in the disputed whale to Bennett as master of the *Massachusetts*. Without a joint property right in the anchored bowhead, Bennett was not able to alter the *Hillman's* legal ownership of the wayward cetacean. Mateship merely gave the *Massachusetts* a possible future right to share in the proceeds from the *Hillman's* catch should – at the end of a designated period – the latter have more bone and oil in its hold. While Sprague's legal analysis of mateship was certainly correct, he failed – as in much of his decision – to see the issue from the whaleman's perspective. Mateship, for whalemen, was about more than legal rights. It was a reflection of a deeper commitment to the shared welfare of two crews. It was also a relationship that merged the interests of the mated ships in the eyes of other captains in the fishery. This blending of the *Hillman* and the *Massachusetts* helps explain why Captains Holt, Blackmer, and Tooker ruled at arbitration that Parker was entitled – based on Bennett's representation that no whale had been taken – to the bowhead anchored by Leonard and the crews of the mated vessels.

V

One of the debates that has animated legal scholarship over the last eight or so decades has concerned the concepts of instrumentalism and formalism in describing and understanding how judges decide cases. Starting in the 1920s, the so-called Legal Realists attacked what they saw as the prevailing formalism of a deeply conservative American judiciary. Legal Realists such as Karl Llewellyn believed that the ability of the law to accommodate the rapid rate of social and technological change of the twentieth century was hindered by a slavish adherence to the past. Nineteenth-century jurists and, in the estimation of the Legal Realists, many of their heirs on the bench believed that cases must be decided by a strict application of rules, precedent, and natural law. When confronted with a novel situation such judges – decried as formalists – relied on deductive reasoning to apply these accepted sources of the law to resolve the issue at bar. The result of this almost mechanistic process of deciding cases was that such judges never considered the effect of their rulings on how people conducted their business. As an antidote to such an inherently conservative way of looking at law, the Legal Realists prescribed what later scholars have termed instrumentalism. Rather than look exclusively to existing laws and cases for guidance, the instrumentalist view was much more expansive in what it was willing to consider as relevant in making the proper decision. Taking the law in the direction of a desired social policy was, for an instrumentalist, far more important than following an established rule. Instrumentalism was, therefore, directed at ends rather than means. For legal scholars of a progressive bent anxious to defend the New Deal, the appeal of instrumentalism was obvious.[32]

Legal historians have largely accepted the formalist/instrumentalist dichotomy as useful in understanding the development of American law in the long nineteenth century. Most notably, Morton Horwitz has explained that in the wake of the Revolution American judges threw off the formalist mantle of English common law and fashioned new, instrumentalist approaches to old issues. While the moment was ripe for an American law that would make good on some of the promises of the new nation's foundational documents, the instrumentalist ends served were, instead, those of the rising class of entrepreneurs and businessmen. In the first half of the nineteenth century, judges, in Horwitz's central example, reconceived the *sine qua non* of a valid contract law from an equitable exchange to the existence – regardless of the advantages in bargaining power enjoyed by one side over the other – of agreed upon terms. Having banished the old common law concern for fairness, judges were able, by mid-century, to push American commercial contracts toward a standard that interpreted agreements to reflect the preferred practices of the business class. Horwitz's narrative reaches an apotheosis of sorts with a discussion of how the established principle that an employer is responsible for torts committed by his workers (*respondeat superior*) was recast to exclude other employees. What Horwitz terms "The Triumph of Contract" explains how judges were able to ignore an employer's duty of care and declare that workers calculate risk of injury in negotiating their wages and terms of employment. Once the capitalist legal agenda was achieved and the impetus for change declined, the American judiciary, in Horwitz's narrative, once again placed precedent and settled rules atop the hierarchy of legal virtues and entered a new period of formalism.[33]

Horwitz's story of nineteenth-century American legal history is, in large part, driven by a conspiracy between judges and a commercial and business elite. While Stephen Presser's comment that the world of Morton Horwitz is "Dark and Doestoyevskyan" is certainly overstated, there is in Horwitz's *The Transformation of American Law, 1780–1860* much about the alleged alliance that is shadowy and mysterious. Horwitz never really explains how it worked.

[32] Useful introductions to the literature on formalism and instrumentalism can be found in Lynda Sharp Paine, "Instrumentalism v. Formalism: Dissolving the Dichotomy," *Wisconsin Law Review* 1978 (1978): 997–1028; Steven M. Quevedo, "Formalist and Instrumentalist Legal Reasoning and Legal Theory," *California Law Review* 73 (January 1985): 119–57; and Keith Swisher, "The Unethical Judicial Ethics of Instrumentalism and Detachment in American Legal Thought," *Willamette Law Review* 43 (Summer 2007): 577–91.

[33] Morton J. Horwitz, *The Transformation of American Law, 1780–1860* (Cambridge: Harvard University Press, 1981). For an appreciation of the extent to which the work of Morton Horwitz has shaped and continues to shape this debate see the following symposium editions dedicated to *Transformation of American Law: Law & Social Inquiry* 28 (Fall 2003): 1121–60 and *Tulsa Law Review* 37 (Summer 2002): 849–965.

How did judges know exactly what merchants wanted? The case of *Taber v. Jenny* is instructive and – while certainly not disproving Horwitz's thesis – suggests that it functions better on the macro than the micro level. The lawyers and the judge in *Taber* never understood what the whalemen wanted from the court. In addition to the obvious – a favorable decision and in the case of Taber and Company perhaps recognition that Parker had fixed the arbitration – both sides sought a tribunal that understood what was important to men who financed whalers and whose economic survival depended upon successful voyages. Whalemen did not care about property rights in the abstract and they certainly were not interested in a decision that would mean fewer whales rendered.[34]

If nineteenth-century courts were so attuned to the needs of commercial interests, why did the judge in *Taber* and, as we shall see, in the other American whaling cases so misunderstand an industry of such importance to the national economy? The first and perhaps simplest explanation is that Judge Sprague, instead, knowingly ignored how the litigants viewed the dispute. Courts found whaling property cases insignificant and Sprague – aware that *Taber* was a case of first impression – had the freedom to view the matter as he saw fit. While insurance and employment issues involving the whaling industry were often litigated, whalemen had proven adept at resolving disputes over contested whales on their own without violence or disruption in the valuable supply of oil and bone. This ability of whalemen to settle arguments without filing actions might have suggested to Judge Sprague that it was better to view the case in terms of general property law rather than the particularities of industry practice. Why issue a ruling on a minor matter applicable only to whaling that might be used as precedent in other situations that could disrupt settled principles of commercial law? Seen in this way, Judge Sprague may have understood that his application of salvage law and definition of a drifting whale was antithetical to whaling custom, but determined that it was a safer basis upon which to draft an opinion.

This possibility is undercut by the fact Sprague's successor on the District Court bench, John Lowell, explicitly stated that it was the singular nature of whaling and the degree to which its crews formed a discrete branch of maritime service that dispelled any concern that precedent in a whaling property dispute would be applied to another form of commerce. The unique nature of whaling reduced its property law, in the view of Lowell and others, to a sort of legal backwater. Sprague even indicated in his *Taber* opinion that he would follow whaling custom if he found it applicable to the facts at bar. It is, therefore, more likely that Judge Sprague did not appreciate how preserving the rights of a slayer to an anchored whale increased the chances that a cetacean would escape

34 Stephen B. Presser, "Revising the Conservative Tradition: Towards a New American Legal History," *New York University Law Review* 52 (June 1977): 700.

the blubber knife and was led, in any event, to believe from counsels' arguments that such a concern was not relevant. The central legal argument offered by Eliot and Pitman on behalf of Taber was, after all, based on *The Institutes of Justinian*.[35]

What ultimately explains this disconnect between whalemen and the legal system was that the latter did not grasp that for the former the taking of the maximum amount of oil and bone while abiding by the laws of honor was more important than a consistent application of property law principles or even such whalemen created customs as fast-fish, loose-fish or iron holds the whale. The entire dispute began with individual decisions by Leonard and Cook to provide false intelligence concerning their success and the location of whales. Had Leonard and Cook not broken the laws of honor, Parker would have likely turned the bowhead back to the *Hillman* which all whalemen would have considered the rightful owner. Likewise, if Parker had not attempted to hide the amount of oil produced by the bowhead and not tried to fix the arbitration hearing, Taber would not have filed the cause of action. Had this dispute been simply a good faith disagreement about which vessel should be granted the whale, it would have been resolved at sea or, as it almost was, in arbitration. As the positions taken by Holt and Tooker in their arbitration deliberations indicate, that resolution would have had more to do with common sense and a loose application of customs than what Melville and subsequent legal scholars have deemed the universal laws of whaling.

What *Taber* suggests is that reading court decisions can be misleading. It is easy to assume that judges and attorneys understood litigants. Without a close examination of what participants considered important, the perception that a judicial ruling was favorable to commercial interests may prove illusory or merely fortuitous. In the case of whaling litigation, the failure to consider the cases from the perspective of men in the Sea of Okhotsk and New Bedford counting houses has led to the incorrect conclusion that courts ratified the customs of whalemen and rendered decisions that maximized the welfare of the industry. The decision in *Taber* – if followed at sea – would certainly not have increased the take of whales.

[35] Judge John Lowell upheld a whaling custom with the observation that whaling "is the only branch of industry of any importance in which it is likely to be much used; and if a usage is found to prevail generally in that business, it will not be open to the objection that it is likely to disturb the general understanding of mankind by the interposition of an arbitrary exception." *Swift v. Gifford*, 23 F. Cas. 559–60 (US District Court, D. MA, 1872).

6

Halves

In *Taber*, the court insisted that an anchored whale bearing the clear marks of appropriation remained the property of its slayer regardless of how long – as in the law of salvage – it remained unclaimed by the owner. The United States District Court in Boston returned to a similar fact pattern ten years later in *Bartlett v. Budd*. In that matter, the court upheld the ruling in *Taber* and again rejected the idea that whaling custom could supersede the established legal principles of how one comes to have and retain ownership of property. While it was clear that in the eyes of the law that the custom awarding a dead whale to the ship that finds and cuts in the animal's flesh before the slayer asserts ownership did not apply to properly marked and anchored whales, the Boston District Court would revisit this practice in a later case. In *Swift v. Gifford*, District Judge John Lowell and counsel extended the cutting in custom to living whales pursued by a first striker and the ultimate slayer. In so doing, the court created out of whole cloth a law of whaling that did not exist at sea. That the hornbook property law of whaling came quickly to reflect the ruling in *Swift* never caused whalemen to alter their practice of exclusively using cutting in as a timeliness test for drifting carcasses. The *Swift* ruling took the one whaling custom that was well understood by whalemen in the Sea of Okhotsk and stretched it to apply to a situation – the chase of living whales – that was never covered by an articulable norm upon which all hunters could agree. As the testimony in the Hawaiian Supreme Court case of *Heppingstone v. Mammen* demonstrated, whalemen in the Sea of Okhotsk held many disparate ideas about which ship should take a living whale pursued by more than one vessel. Legal scholars, relying on the *Swift* opinion, have mistakenly seen in whaling custom an example of how well settled norms that were created by participants come to be adopted by courts. In fact, American whalemen played a limited role in creating the law of whaling as stated in legal treatises. At sea and during negotiations in port, inchoate ideas of fairness and a desire to avoid litigation and maintain order often resulted in the suggestion that a whale be split.

I

The first week of July 1856 found the Okhotsk fleet in its customary early summer position awaiting the breakup of the ice blocking entrance to the Southwest Bay. Having departed New Bedford in the Autumn of 1853, the *Canton Packet* was beginning its third season in the North Pacific. First Mate Hiram Norton spent several days in his boat probing the ice for an opening along the coast to the fertile summer bowhead grounds near Shantar Island. The usual urgency of finding a way through the ice was surely heightened by the late degradation of the ice that season. In some years, the ice melt permitted access to the Southwest Bay as early as the penultimate week of June. The experienced Norton recognized that the outgoing tide pushed the ice away from shore allowing a boat to use these temporary paths to push southward. On July 14th, Norton's boat and one under the command of Edward Cook, the *Canton Packet's* second mate, passed through a belt of ice and progressed a short distance into the Southwest Bay before the ice, riding the incoming tide, forced them to seek refuge on the beach. Norton and his shipmates took advantage of their good fortune as some of the first boats into Southwest Bay and managed – in the brief absence of significant competition – to kill four small whales the following day. The success of the *Canton Packet's* boats raised an issue familiar to whalemen plying the bays of the Sea of Okhotsk. Having squeezed through narrow holes in the ice, boats sometimes waited days or even weeks before their ships could join them in open water. Captured whales were typically secured along the beach until they could be taken shipboard for rendering. On occasion, however, the towing of a dead whale to shore was precluded by the tide, weather conditions, or other exigencies. Whales, in such situations, were anchored where they were killed to wait towing in more favorable conditions.[1]

[1] *Bartlett v. Budd*, 2 F. Cas. 966 (US District Court, D. MA 1868). Depositions of Gilbert Borden, Hiram Norton, and Gamaliel Fisher, Case File, *Bartlett v. Budd*, United States District Court, Massachusetts, March, 1868, United States National Archives, Northeast Region, Waltham, Massachusetts (National Archives, Northeast Region). For a description of when and where whales were hunted in the Sea of Okhotsk, see, Otto Lindholm, "Whales and How Tides and Currents in the Sea of Okhotsk Affect Them," January 1, 1965, Scripps Institution of Oceanography Technical Report, accessed September 25, 2015, http://escholarship.org/uc/item/8j93s3pv; Kim E. W. Shelden and David J. Rugh, "The Bowhead Whale, Balaena Mysticetus: Its Historic and Current Status," *Marine Fisheries Review* 57, issue 3/4 (1995): 1–20; Sue E. Moore and Randall R. Reeves, "Distribution and Movement," in *The Bowhead Whale*, eds. John J. Burns, J. Jerome Montague, and Cleveland J. Cowles (Lawrence, KS: The Society for Marine Mammalogy, 1993), 357–60; C. M. Scammon and Edward D. Cope, "On the Cetaceans of the Western Coast of North America, *Proceedings of the Academy of Natural Sciences of Philadelphia* 21 (1869): 13–63; and Charles M. Scammon, *The Marine Mammals of the North-western Coast of North America* (New York: Dover Publications, Inc, 1968), 52–65. Making the proper connection

The *Canton Packet*, having quickly followed its boats through the ice, was nearby and Norton had the luxury of directing Cook to tow three of the whales back to the ship. Norton was forced – for reasons not stated in his deposition – to anchor the fourth whale in the bay before retiring to the beach to sleep. The next morning, Norton killed a whale in five or six fathoms of water about four to five miles from the shore. The small bowhead – or poggy as whalemen termed young cetaceans – began to sink which, in combination with an unfavorable tide, convinced Norton that towing his catch to shore or the ship was for the moment impractical. Having used his lone anchor the day before, Norton was without means to secure his latest catch. Fortunately, Gamaliel Fisher, a townsmen of Norton and first mate of the *Barnstable*, was in the vicinity and loaned an anchor to Norton. The anchor, bearing the *Barnstable's* name, was attached to the whale by a rope over thirty fathoms in length. To aid in future recovery, Norton buoyed the whale with his main sail, an oar, and a paddle. Anchored whales, as Norton later testified, are apt to sink out of view with the action of the tides. Norton's claim to the whale was further marked by harpoons from the *Canton Packet* and the *Barnstable* affixed to the animal's back. Fisher, who in addition to his fortuitous arrival and loan of an anchor and iron provided evidence corroborating Norton's testimony, averred that – in his opinion – the whale was securely anchored. Confident that the whale would not drift in the calm conditions and noting landmarks on shore to fix the location, Norton and his crew departed for a night on the beach.[2]

The following morning, Norton set out – in company with Fisher's boat – to relocate his anchored whale. As the boat drew near the anchorage, Norton observed three boats belonging to the *Emerald* of Sag Harbor approaching from where he had left the whale. The *Emerald's* boats were in the process of hauling up an anchor attached to a dead whale. Norton drew near and declared that the whale belonged to the *Canton Packet*. The *Emerald's* mate denied the claim and responded to Norton's questions by asserting that they had killed the whale the previous day further up the bay. Norton protested that this was a lie and requested that the whale be turned over to reveal the harpoons of the *Canton Packet* and the *Barnstable*. The *Emerald's* mate ignored Norton's demands and the confrontation lasted only ten or fifteen minutes before Norton left to locate his other anchored whale. Both sides to the dispute were well aware that the *Emerald's* story of killing the whale in question was, as Norton

between what whalemen called an area and what appears on a modern map is challenging. Whalemen were not always consistent and the transliteration of Russian names adds to the confusion. For a reprint of an 1852 logbook sketch of the Sea of Okhotsk labeled with place names used by whalemen, see Randall R. Reeves, Tim D. Smith, and Elizabeth A. Josephson, "Observations of Western Gray Whales by Ship-Based Whalers in the Nineteenth Century," accessed September 25, 2015, www.worldwhalinghistory.org/documentsavailableonline.

[2] Depositions of Hiram Norton and Gamaliel Fisher, Case File, *Bartlett v. Budd*, National Archives, Northeast Region.

alleged, false. Norton, nevertheless, had no real option other than retreat with the hope that the captain of the *Canton Packet* would take up the issue directly with the *Emerald's* master. Norton could, of course, have attempted to take the whale by force. Outnumbered three boats to one and apparently not counting on the assistance of Fisher who was still in the vicinity, Norton explained in his deposition that "if I had boats enough I shouldn't have let him had it." Norton's calculation of the force he would need to retake the whale was not unique. Participants in such confrontations often indicated that their lack of numbers precluded a self-help remedy at sea.[3]

Yet, if a simple head count was enough to prevent violence, why did armed struggle not break out when the strength of disputants was even or one or both sides miscalculated their relative strength? Some crews must have looked across a whale and decided that their opponent's boat was filled with men of smaller stature or perhaps of advanced age who could be easily overpowered. The absence of any evidence of violence at sea over a contested whale suggests that something more than a simple weighing of the prospects for success was at work. Whalemen were not as a group, after all, distinguished by their gentility. There were many incidents of violence aboard ship and in various ports of call. The likely answer can be found in the rigid social and command structure of nineteenth-century whaling. Operating on a military model, captains exercised – in theory and often in practice – near absolute authority over their officers and crews. While masters had different ideas as to the sorts of issues requiring a strong show of authority, captains were certainly more likely to assert their authority in matters concerning the ultimate success of the voyage. Petty disputes between crew members that might ultimately escalate into a violent – often alcohol fueled – confrontation were of less immediate concern to captains and individuals were left to their own means of settling a score. Theft from the ship's stores and desertion were, by contrast, matters of greater concern to captains. No issue implicated a voyage's success and a captain's standing in the community of his peers as much as a conflict with another ship over a whale. In these situations, officers who commanded the boats on the front lines of ownership disputes were well aware that they must act in accordance with their captain's orders. It is unlikely that these officers were authorized to initiate a physical altercation with another ship's crew to take a whale. Mates such as Norton could bluster and speak of violence, but they ultimately left resolution of these disputes to their captains.[4]

Having conceded the whale to the *Emerald*, Norton returned to the *Canton Packet*. Norton explained to Captain Gilbert Borden that the *Emerald's* "boats

[3] Depositions of Hiram Norton and Gamaliel Fisher, Case File, *Bartlett v. Budd*, National Archives, Northeast Region.

[4] For the structure of whaling communities and the types of issues of greatest concern to captains, see Thomas Stone, "Atomistic Order and Frontier Violence: Miners and Whalemen in the Nineteenth Century Yukon," *Ethnology* 22, no. 4 (October 1983): 327–39.

have got my whale, and that he had better go right off and get the whale." Borden immediately boarded a boat to demand the whale from the *Emerald*. The weather deteriorated and Borden was unable to reach the *Emerald* against a strong wind and unfavorable tide. Poor conditions persisted and it was several days before Borden could again set forth to state his case. Captain Frederick Hallock of the *Emerald* offered that while he did not doubt Borden's story that his boat had killed the whale and secured it with the *Barnstable's* anchor, the whale, nevertheless, was now the property of his ship. Hallock explained to Borden that his boats had discovered the whale with attached anchor adrift. The whale, Hallock further related, was brought to the *Emerald*, cut in, and boiled – producing a paltry ten to twelve barrels of oil. The basis of Hallock's claim to the prize at sea and in court was the whaling custom that a boat finding a whale may keep it so long as it is cut in prior to the assertion of ownership by the slayer.[5]

The custom Hallock cited to Borden was, of course, the same usage advanced without success by the *Zone* in *Taber v. Jenny*. As the *Taber* decision was rendered a mere six months prior to the dispute between the *Emerald* and the *Canton Packet*, Hallock would have had no way of knowing that Judge Peleg Sprague had ruled that a drifting whale marked with "unequivocal marks of appropriation" remained the property of the slayer. Whether knowledge of Sprague's holding would have altered Hallock's argument or actions at sea is unknowable. Counsel for the *Emerald* in the matter captioned *Bartlett v. Budd* was, however, fully aware of the decision and the unfavorable precedent it set. Accepting the facts as presented by Borden aboard the *Emerald* and repeated by witnesses for the *Canton Packet* at trial, the two matters were remarkably similar. The anchor and irons affixed by Norton was exactly the sort of mark of appropriation that Sprague had in mind. The chance of Judge John Lowell overruling the holding of Sprague – his predecessor whose experience in whaling matters he lauded in a later case – must have appeared very remote to *Emerald* attorney J. C. Stone. Faced with an unfavorable precedent, Stone attempted to show that the facts were not as the witnesses for the *Canton Packet* propounded. If Stone could demonstrate that the whale had not been unequivocally marked as property of the *Canton Packet*, then the custom that awarded a drifting whale to a finder who cut in prior to a claim would decide the case. In *Taber*, Sprague had, after all, indicated that – but for the marks of appropriation – the *Zone* would have prevailed pursuant to the custom of the fishery.[6]

[5] Depositions of Hiram Norton, Gilbert Borden, and Frederick Hallock, Case File, *Bartlett v. Budd*, National Archives, Northeast Region.

[6] *Taber v. Jenny*, 23 F. Cas. 605 (US Dist. Ct., D. MA, 1856). For a full discussion of *Taber*, see Chapter 5. Lowell's praise of Sprague is at *Swift v. Gifford*, 23 F. Cas. 558 (US Dist. Ct., D. MA, 1872). Judge Lowell's reluctance to overrule Sprague's *Taber* holding may – assuming that he had any inclination to do so – have been strengthened by his appreciation that the matter was originally assigned to Judge Sprague. The twelve-year journey from a dispute in the Sea of Okhotsk to a

Emerald first mate John Fowler testified at trial by way of an *ex parte* deposition that he had found the disputed whale adrift in about ten fathoms of water on a foggy morning with no other boats in the vicinity. An anchor bearing the name of the *Barnstable* was attached to the whale by a ten fathom rope. The line was so tangled around the whale and anchor that the latter was not in contact with bottom of the bay. Second mate Herbert White added that although the anchor and line were in good condition, the ten fathom rope was not long enough to hold the whale in water he estimated to be eighteen fathoms in depth. Weather conditions were also such, White added, that the seventy to eighty pound anchor would not have been sufficient to secure a whale in the prevailing currents running at five to nine knots. Both Fowler and White testified that there were no irons affixed to the whale and that they towed their prize to the *Emerald* without encountering any claimants. Captain Hallock also stated that there were no irons in the whale when it was brought to the *Emerald* to be cut in and boiled. While Fowler, White, and Hallock agreed that the applicable custom required that a claim be made prior to cutting in, Fowler added – after initially stating that no usage governed the situation and then correcting himself – that the craft must bear the name of the claimant. As whalemen defined craft to be harpoons or lances, an anchor was not – according to Fowler – sufficient to support a claim to a drifting whale. The *Emerald* presented, therefore, evidence that – if deemed credible – would allow the court to honor the holding in *Taber* while finding that Norton had not marked the whale to a degree sufficient to supersede the custom of the fishery.[7]

ruling in the United States District Court was – while excessive by even the standards of whaling litigation – likely the result of the usual difficulties in deposing witnesses and gathering evidence. What is more difficult to explain is why the firm of Ivory Bartlett, the owner of the *Canton Packet*, filed and pursued over the course of many years an action concerning such a small whale. Oliver Prescott, chosen by Judge Lowell to determine the appropriate monetary judgment, declared the value of the whale at the time of its killing to be $683.52. He also calculated the interest from July 14, 1856 to April 14, 1868 to be $481.86. The total judgment, exclusive of costs, was $1165.68. Case File, *Bartlett v. Budd*, National Archives, Northeast Region. Perhaps, the matter – as in the case of *Taber* – had become personal. The only evidence of this, however, can be found in the deposition of the *Canton Packet's* master. Captain Borden testified that two members of the *Emerald's* crew approached him in Honolulu and indicated that their ship had stolen the whale from the *Canton Packet*. It is also possible that the litigation was part of a purely economic calculation. The firm of Ivory Bartlett was struggling – like most whaling concerns – in February 1858 when the complaint was filed. In January 1858, Dun and Company reported the value of the firm at $40,000 and commented "Has prob. been doing a losing bus the last yr but has not gone astearn eno. to embarrass him." By 1867, Ivory Bartlett was moving out of whaling and thought to be worth $150,000 which was mostly in stocks. Massachusetts, Vol. 17, p. 450, R.G. Dun & Co. Credit Report Volumes, Baker Library, Harvard Business School.

7 Depositions of John Fowler, Herbert White, and Frederick Hallock, Case File, *Bartlett v. Budd*, National Archives, Northeast Region. For a discussion of the word craft, see James Temple Brown, *The Whale Fishery and Its Appliances* (Washington, D. C.: Government Printing Office, 1883), 6. "The term 'craft' includes the harpoons, lances, boat-spade and boat-hook, but is oftentimes more specifically applied to implements used to strike and kill the whale."

The legal argument presented by T. M. Stetson on behalf of the *Canton Packet* was vigorous and attacked the *Emerald's* position on several fronts. Stetson, hoping to effectively guarantee that his version of the facts would prevail, asserted that the testimony of Fowler, White, and Hallock was inadmissible. While the near impossibility of securing attendance at trial from a particularly peripatetic group of witnesses necessitated the use of depositions in whaling cases, the general practice was for attorneys for both parties to be present when individuals – back in New England between voyages – were deposed. This allowed for the adversarial examination and cross examination essential to the fact discerning process in a common law proceeding. The captain and first and second mates of the Sag Harbor based *Emerald* were all residents of New York and did not come to Massachusetts to be questioned. Instead, Stone sent – with court approval – depositions to New York to be administered by a United States Commissioner, Notary Public, or Justice of the Peace. For unknown reasons, Stetson either chose not submit his own questions or was not afforded the opportunity to at least replicate in some measure the adversarial process. Counsel cited for Judge Lowell's consideration a number of cases concerning the admissibility of *ex parte* depositions in cases at Admiralty. Lowell did not explain his ruling, but in allowing the depositions he likely read the cited authorities to permit admission "with the greatest caution."[8]

Having failed to keep the *Emerald's* evidence out of court, Stetson argued that the custom governing rights in a drifting whale was inadmissible. Stetson cited a line of cases that limited the use of a custom or usage of a trade to the interpretation of uncertain terms or clauses in contracts between the parties. Customs cannot, according to these cases, be used to contradict the clear meaning of a contract. In *The Brig George*, a First Circuit Court opinion cited by Stetson, Justice Joseph Story stated that even when general, the customs of ports or other locations are not admissible "to control or alter the settled maritime law." If Stetson could convince Lowell that the custom advanced on behalf of the *Emerald* was inadmissible, the matter would be governed by general principles of the law of pursuit and capture of whales as *ferae naturae*. Pursuant to Sprague's understanding in *Taber* of ownership, a whale that is killed becomes every bit as much the property of the slayer as his harpoons. Assuming *arguendo* that the custom was admissible, Stetson asserted that the witnesses for the *Emerald* had misstated the usage by eliding two important

[8] *American Insurance Company v. Johnson*, 1 Blatchford & Howland 9, 30 (1827). The court's case file does not contain legal briefs filed by counsel, but it does include notes of what appears to be the cases cited by Stetson and Stone during oral argument. While it is impossible to know who took these notes, they were likely made by Judge Lowell during an oral argument as a slightly annotated list of cases to consult in rendering his decision. Case File, *Bartlett v. Budd*, National Archives, Northeast Region. Trial courts – particularly in matters without juries – will, in the author's experience, often allow evidence favorable to the losing party to be admitted as a means of forestalling a potential issue on appeal.

exceptions to its application. Witnesses for the *Canton Packet* – whose identity and testimony have not survived – revealed, according to Lowell's opinion that the custom only applied to whales without marks of appropriation. Harpoons or lances did not, in this explanation of the usage, constitute a sign of appropriation as such craft were frequently attached to whales that escaped without suffering any real injury and before a whalemen could exercise any control over the animal. The second exception to the custom was that it only applied to boats that were in waters deeper than 100 fathoms or, in the nautical vernacular, off soundings.[9]

Judge Lowell clearly indicated in the first sentence of his opinion that his decision awarding the whale to the *Canton Packet* was not based on the customs of whaling. "A whale, being *ferae naturae*, does not," Lowell explained, "become property until a firm possession has been established in the taker." Echoing a long tradition of animal capture cases dating back to Justinian, Lowell – citing *Taber* – stated that once the *Canton Packet* firmly and completely possessed the whale, it had the same right of ownership in the cetacean as it had in the attached anchor. While explicitly dodging the issue of the power of a usage to alter "the strict law of the pursuit and capture of whales and their appropriation," Lowell opined that the custom, as presented by the *Canton Packet*, was supported by the preponderance of the evidence. Whales in the Sea of Okhotsk marked with an anchor or some other sign that it had come under the control of the slayer were not free for the taking. Lowell went on to point out that he would not uphold a custom such as the one propounded by the *Emerald* that allowed an opportunistic interloper to swoop in and grab what belonged to another ship. Custom, as employed by Judge Lowell, supported the law of *ferae naturae* by supplying guidance to whalemen as to what constituted capture in the very particular context of whaling.[10]

[9] *The Brig George*, 1 Sumner 151, 158 (1832). For a definition and discussion of soundings and nineteenth-century techniques of determining water depth, see *A Naval Encyclopedia: Comprising a Dictionary of Nautical Words and Phrases; Biographical Notices, and Records of Naval Officers; Special Articles on Naval Art and Science, Written Expressly for this Work by Officers and Others of Recognized Authority in the Branches Treated by Them. Together with Descriptions of the Principal Naval Stations and Seaports of the World* (Philadelphia: L. R. Hamersly & Co., 1881), 144–5, 198–212, 616. The legal arguments presented in *Bartlett* can be pieced together by reading Judge Lowell's opinion and the notes in the case file discussed in the preceding footnote. *Bartlett v. Budd*, 2 F. Cas. 966 (Dist. Ct., D. MA, 1868) and John Lowell, *Judgments Delivered in the Courts of the United States for the District of Massachusetts* (Boston: Little, Brown, and Company, 1872), 223–5. For some of the most relevant cases and authorities cited by counsel concerning the use of customs, see *The Mutual Safety Insurance Company v. Hone*, 2 NY 235 (1849); *The Schooner Reeside*, 2 Sumner 567 (1837); *Two Hundred and Ten Barrels of Oil*, 1 Sprague 91 (1844); *Post v. Jones*, 60 US 150 (1856); and James Kent, *Commentaries on American Law* (Philadelphia: The Blackstone Publishing Co., 1889), 3:section 260.

[10] *Bartlett v. Budd*, 2 F. Cas. 966.

The disputes that came to be litigated as *Taber* and *Bartlett* involved whales that had been killed, anchored in the water, and left unattended until they could be towed to the ship for rendering. Such situations were possible only in waters that were shallow enough to permit anchorage or featured large pieces of ice to which a whale could be appended. In the Greenland fishery, the early system of assigning bays by nationality or allowing *de facto* division prevented contests over whales anchored in shallow water. There was no question as to whom such whales belonged. To take a whale anchored in a Spitzbergen bay was clearly viewed as theft. When Greenland whalemen ventured into deeper water, Dutch statutory law and the custom of fast-fish, loose-fish determined that ownership of a dead whale could not be preserved by securing it to ice. Such a whale was deemed a loose-fish unless a crew member was left behind to guarantee later recovery. In the bays of the Southern fishery, the absence of ice and the more favorable weather conditions greatly reduced the necessity of anchoring whales for later recovery. When whaling shifted to the bays of the Sea of Okhotsk with their treacherous weather conditions and with the rise of the practice of sending boats out for days or weeks to hunt, the anchoring of whales became a standard feature of the grounds. In a fishery that generally honored the doctrine of iron holds the whale, the practice of anchoring whales was problematic. As with any application of iron holds the whale, it was challenging to create a test that would protect the interests of the first striker and prevent whales from going unused. Whalemen were, after all, not inclined to see a whale go to waste when the slayer was not in the immediate vicinity. The notoriously strong and variable currents in the Sea of Okhotsk served to exacerbate the concern that an anchored whale might never be recovered by its slayer. The decisions in *Taber* and *Bartlett* provided a ready solution to the problem. A dead whale secured with a marked anchor remained the property of the slayer. While whalemen may not have liked that such a prize might drift off and escape utilization, judges and lawyers – trained to view ownership as fully vesting when a *ferae naturae* is reduced to complete possession – were not willing to countenance an approach which denigrated the certainty and sanctity of what it means to own something.[11]

[11] For the customs in Greenland concerning anchored whales, see Chapter 3. The practices of the Southern fishery are discussed in Chapter 4. On at least one occasion a dispute over an anchored whale that broke free and was claimed by another party was litigated in Australia. In 1842, poor weather forced the anchoring of a whale in a bay on the coast of Victoria. The whale broke free and was discovered by two men on a nearby beach. The finders, having begun to render the whale, were not willing to accept a portion of the oil offered by the slayers. A magistrate found that the whale clearly belonged to the slayers who had anchored their catch. This matter was reported in the *Portland Bay Guardian and Normanby General Advertiser* of August 20, 1842 and reprinted on September 14, 1842 in the *New South Wales Examiner*. This information was provided by whaling historian Mark Howard of Melbourne, Australia in correspondence dated August 18, 2008.

II

The other three cases involving American ships litigated in the nineteenth century concerned whales that were struck by one whaler and then killed by the crew of another ship. In what might be termed the classic whale dispute scenario, the first striker affixed a harpoon and remained in pursuit of the whale when the second boat killed and took possession of the animal with the initial iron still attached. The 1872 case of *Swift v. Gifford* has been viewed by legal scholars as confirmation of the universality of iron holds the whale in the North Pacific and that the custom of allowing such claims to be made until cutting in was adopted to govern the troublesome issue of timeliness. A close reading of the case file reveals that the confusing way in which Judge Lowell stated the custom of whaling in his opinion has contributed to the mistaken view that norms in the Sea of Okhotsk were well settled. As the trial testimony in the Hawaiian case of *Heppingstone v. Mammen*, however, makes clear, whalemen in the Sea of Okhotsk were deeply conflicted as to the applicable standard for awarding disputed whales. Instead, negotiation based on a vague sense of the right way to handle such disputes – not the clear guidance of time honored customs – was employed to keep contests out of court.

At about 9:00 PM on the evening of August 1, 1867 George Silvey – second mate of the *Hercules* – was below deck when the call went out to lower the boats. It was still light in the North East Gulf of the Sea of Okhotsk as the crew scrambled to pursue a spouting whale spied moving quickly in the direction of the *Hercules*. The boat commanded by Silvey managed to cut the whale off, but the bowhead sounded and remained under water for the next twenty to thirty minutes. When the whale surfaced at a distance of about a half-mile, Silvey was able to quickly close the gap and affix a harpoon. The injured bowhead ran off with iron and rope still attached. It was at a distance of more than two miles and sixty to ninety minutes before Silvey's boat hauled up close enough to – with the assistance of other boats from the *Hercules* – administer the fatal lance jabs. As the whale neared death, Captain Nehemiah Baker of the *Rainbow* approached in a boat and claimed the bowhead. Baker explained that the *Rainbow* had struck the whale a short time before and that it had escaped with the affixed harpoon and about two hundred fathoms of line. Although Silvey had noticed an additional rope about the whale, he had not realized that it was connected to a harpoon. The whale, with death, rolled up revealing an iron. Removed at Baker's request, the harpoon bore the *Rainbow's* mark.[12]

[12] Depositions of George Silvey and Robert D. Gifford, Case File, *Swift v. Gifford*, United States District Court, Massachusetts, March, 1872, United States National Archives, Northeast Region, Waltham, Massachusetts (National Archives, Northeast Region). Silvey's name is also spelled Silvia and Sylvia. Silvey has been used throughout as that is the spelling used by the notary who took his deposition. The material in the case file from which to reconstruct the facts in the dispute is limited to the depositions of Silvey and Gifford and the pleadings. As Silvey and Gifford were both mates on the *Hercules*, the *Rainbow's* version of the events must be gleaned

The negotiations that attended so many confrontations at sea over whales commenced with Baker's claim and *Hercules* first mate George Manchester declining to state a position. Manchester indicated that he would leave the discussion to his captain, Isaac C. Howland, who was on board the nearby *Hercules*. It was agreed that Baker would take possession of the whale while Howland was consulted. Dissatisfied with this arrangement, Silvey was not reluctant to express his opinion as to the proper disposition of the whale. He argued with Baker and perhaps his own first mate that the whale should be evenly divided between the ships. When Captain Howland arrived on the scene about a half hour later, he took up Silvey's position. Howland explained to Baker that his officers claimed one half of the whale as the *Hercules* had foregone pursuit of another whale in taking the subject of the present dispute. In addition, Howland asserted, the *Rainbow* would not have been able to catch up with the whale but for the intervention of the *Hercules*. The *Hercules* had, in effect, done Baker a favor in returning to him his iron and line which would otherwise have been lost in the side of a whale quickly moving into the protection of a darkening sky. Baker was apparently not moved by Howland's argument and the captains agreed to let their owners decide the matter at a later date.[13]

The owners of the *Hercules*, the firm of Swift & Perry, were obviously not satisfied by their discussions with Charles Gifford, the principal investor in the *Rainbow*. Wishing perhaps to expedite resolution of the dispute, Swift & Perry filed an action on August 3, 1870 in United States District Court in Boston, nearly eight months prior to the return of the *Hercules* from the Pacific. In legal argument at trial, John Dodge – counsel for Swift & Perry's *Hercules* – asserted that the matter must be decided by an application of the common law principles of *ferae naturae*. Citing *Buster v. Newkirk* and *Pierson v. Post*, Dodge maintained that wild animals are reduced to ownership only when they are in the total control and possession of a claimant. Dodge was well aware that the customs of whalemen did not require a degree of possession necessary at common law to constitute ownership of a *ferae naturae*. His law partner, Charles Bonney, had, in fact, signed a stipulation that competent witnesses would testify that the uniform usage of New Bedford and Nantucket whalemen had long been that a whale escaping with an affixed harpoon belongs to the first striker so long as he can make a claim prior to the cutting in of the ultimate slayer. It is not entirely clear why Bonney – presumably with Dodge's concurrence – would admit to a custom which would, if accepted by the court as determinative, doom his client's case. Given the facts in the case, Dodge and

from its answer to the libel. Pleadings are not always particularly helpful in determining what happened at sea. In whaling cases they were frequently drawn up by attorneys after discussing the matter with owners who were not present at the dispute. In some cases, such as the libel in *Swift v. Gifford*, pleadings were filed prior to return of a vessel to Massachusetts.

13 Depositions of George Silvey and Robert D. Gifford, Case File, *Swift v. Gifford*, National Archives, Northeast Region.

Bonney might have felt that any statement of whaling custom would result in a negative ruling. Bonney, present at a deposition of veteran whaling captain Abraham W. Pierce, might have, therefore, been willing to concede to any version of the custom. Trial attorneys will occasionally stipulate to unfavorable testimony as a way of mitigating the harm and potential for further damage presented by a persuasive witness. Bonney did not even bother to subject Captain Pierce to cross examination at the time of his deposition.[14]

The decision of Dodge and Bonney to rest the case for the *Hercules* on Judge Lowell ignoring custom and honoring common law was not as ill advised as it might seem. Dodge argued in court before Lowell and in a surviving brief that the custom to which he conceded was void as contrary to the clear statement of the common law. The authority for the position was impressive. In 1837, Joseph Story, a Supreme Court justice and a towering figure in Massachusetts legal circles, wrote in *The Schooner Reeside* of his dissatisfaction with "the almost indiscriminate habit, of late years, of setting up particular usages or customs in almost all kinds of business and trade, to control, vary, or annul the general liabilities of parties under the common law, as well as under the commercial law." The danger of admitting such customs was the confusion and the potential for abuse that would be created. Story explained that the proper use of custom was "to interpret the otherwise indeterminate intentions of parties, and to ascertain the nature and extent of their contracts, arising not from express stipulations, but from mere implications and presumptions, and acts of a doubtful or equivocal character." In *Bartlett*, Judge Lowell had, in addition, explicitly left open the question of the degree to which a usage could alter "the strict law of the pursuit and capture of whales." In honoring the rights of one who killed and anchored a whale, Lowell was accepting a custom that was very much in line with Roman and common law conceptions of how one comes to own a wild animal.[15]

Reading the opening lines of Lowell's opinion in *Swift*, counsel for the *Hercules* might well have judged their reliance on common law a successful strategy. Lowell began his opinion with the statement that common law – borrowing from Roman law – defined ownership of a wild animal as vesting only with "actual and complete possession." The judge then noted the confusion in what he termed the "modern civil law" as to whether "fresh pursuit" combined with a reasonable prospect of ultimately capturing the animal was

[14] Complaint, Stipulation, and Deposition of Abraham W. Pierce. Case File, *Swift v. Gifford*, National Archives, Northeast Region. *Buster v. Newkirk*, 20 Johns. 75 (1822) and *Pierson v. Post*, 3 Cai. 175 (1805). For a discussion of *Buster*, see text accompanying footnote 30 in this chapter. The literature on *Pierson* is enormous. For some of the most recent scholarship, see Bethany R. Berger, "It's Not About the Fox: The Untold Story of *Pierson v. Post*," *Duke Law Journal* 55 (April 2006): 1089–1143; Andrea McDowell, "Legal Fictions in *Pierson v. Post*," *Michigan Law Review* 105 (February 2007): 735–77; and Angela Fernandez, "The Lost Record of *Pierson v. Post*, the Famous Fox Case," *Law and History Review* 27 (Spring 2009): 149–78.

[15] *The Schooner Reeside*, 2 Sumner 567, 569 (1837). *Bartlett v. Budd*, 2 F. Cas. 966.

sufficient. If Lowell followed the stricter interpretation of *ferae naturae*, the *Hercules* would clearly prevail. If, on the other hand, Lowell believed that the case must turn on the likelihood that the *Rainbow* would – without the actions of its competitor – have taken the whale, the *Hercules* produced compelling evidence that without its intervention the animal would have escaped. Silvey testified that the *Rainbow* was at a distance of about eight miles when he struck the whale. Even Gifford conceded in his answer to the complaint that the *Rainbow*, while still in pursuit, was about six miles from Silvey's boat. That Dodge and Bonney directed their argument and testimony at this issue was confirmed by the editor of the reported opinion who observed that the only issue contested at trial concerned the prospects of the *Rainbow* catching up to the fleeing whale.[16]

Instead of applying the common law, Lowell, however, abruptly turned to the customs of whaling and – in the process – made a remarkable admission. Lowell revealed that he did not pursue the common law for the simple reason that it would be impossible to determine whether the *Rainbow* "though continuing the chase, had more than a possibility of success." It appears that Lowell abandoned what he suggested to be the governing law for the simple reason that it would be difficult to apply to the facts at bar. Judges are, of course, frequently faced with facts that – when applied to the pertinent standard – do not suggest an easy resolution to the dispute. The proper solution in such a situation would be to find that in the absence of evidence that the *Rainbow* would have caught up to the whale, the animal must be awarded to the *Hercules*. In abandoning the common law, Lowell took note of Story's concern that customs would unsettle established legal principles, but determined that the application of whaling usage posed no such danger. Whaling practices, Lowell explained, were limited to such a discreet industry and too particular to attract and confuse those engaged in other branches of commerce. Adherence to whaling custom rather than the laws of a particular nation would – as Judge Chambre stated in an oft cited portion of *Fennings v. Lord Grenville* – prevent confrontations at sea between whalemen hailing from around the world.[17]

In turning to custom to decide the case, Lowell did not, however, apply the usage which Bonney had – on behalf of the *Hercules* – agreed would be testified to by competent witnesses. Bonney stipulated that such witnesses would establish that a whale that escapes with the harpoon of a ship and is subsequently killed by another vessel belongs – at least in part – to the first striker if it claims the animal "before it is fully cut in by the second ship." While the stipulation stated that it applied to a live whale pursued by two ships, Lowell correctly recognized that the custom that honored a claim made prior to completion of

[16] *Swift v. Gifford*, 23 F. Cas. 558. *Bartlett v. Budd*, 2 F. Cas. 966. Deposition of George Silvey and Answer, Case File, *Swift v. Gifford*, National Archives, Northeast Region.

[17] *Swift v. Gifford*, 23 F. Cas. 558. *Fennings v. Lord Grenville*, 1 Taunt. 241 (1807).

the cutting in process pertained exclusively to dead whales found adrift in the water. As Lowell explained, a dead whale drifting in the current belongs to the finder so long as it can be cut in prior to a claim being made by the ship whose harpoon the cetacean bears. That Lowell viewed this as a separate, but certainly related, custom was indicated by his use of the word contrary to describe its relationship to the rule that iron holds a living whale. Ignoring the stipulated custom, Lowell, instead, proclaimed the applicable usage for a living whale to be "that he who first strikes it so effectually that the iron remains fast should have the better right, the pursuit still continuing, . . ." The source for this custom was the oral decision rendered by Judge Sprague in the 1863 matter of *Bourne v. Ashley*. While counsel for the *Hercules* argued that the failure of the *Bourne* decision to be published was likely the product of Sprague's dissatisfaction with its contents, the real reason may well have been that it was never reduced to writing. Lowell indicated that his source for the opinion were the notes taken by one of the attorneys in the matter. Having assured the parties that he had examined the *Bourne* case file and found the purported opinion to be consistent with its contents, Lowell determined that the *Rainbow* had maintained pursuit and should be awarded the whale.[18]

The case of *Swift v. Gifford* captures, in many respects, the vagaries of whaling custom and the twisted application of those practices in the courtroom. This disconnect between practices at sea and how they were used in court was demonstrated by the conflation of the different approaches governing live and dead whales. While whalemen harbored many different ideas about what constituted possession of a live whale – as the discussion of *Heppingstone v. Mammen* will demonstrate in the next section – the custom concerning a harpooned dead whale was, by whaling standards, well settled at sea. It was generally understood that a ship which discovered a dead whale bearing the harpoon of another vessel was entitled to as much as it could cut in prior to a claim being made by the first striker. It was this custom that was extended by the expert witness and counsel in *Swift* to living whales. This broadening of the usage concerning dead whales appears novel and marks a clear departure from accepted practices. Even Judge Lowell chose to ignore this standard – in spite of the stipulation of counsel – and followed a version of iron holds the whale that preserved the rights of the first striker so long as they maintained pursuit of their catch. Francis Allyn Olmsted's 1841 account of a whaling trip to the Pacific is typical in limiting cutting in as a test for timeliness to dead whales. As Olmstead revealed, the assumption was that a dead whale had been mortally wounded by the boat that attached a harpoon. The slayer was entitled to the whale pursuant to iron holds the whale, but only if he was able to make his claim before the whale was cut in. This version of iron holds the whale as applied to dead whales protected the slayer who was able to quickly find its

[18] *Swift v. Gifford*, 23 F. Cas. 558. *Bourne v. Ashley*, 3 F. Cas. 1002 (D. MA 1863).

prey and it increased the likelihood of harvesting should the whale – as frequently happened – succumb to its wounds at some distance from the boat that dealt the fatal blows. Research to date has yet to discover a single use of the cutting in test for the timeliness of a claim applied to a living whale pursued by two ships.[19]

The decision of counsel for the *Hercules* to agree that competent witnesses would testify that cutting in marked the time within which a first striker must claim a whale killed by another vessel was, as previously indicated, likely made with little attention to the particulars of the stipulated custom. The strategy of Dodge and Bonney was, instead, to convince Judge Lowell that he must follow the common law of *ferae naturae* and not custom in rendering a decision. It is also probable that presented with a single expert witness who would testify as to the custom, counsel – believing Captain Pierce's testimony as to the universality of this usage – saw no grounds for challenging his expertise. In accepting Pierce's testimony, Dodge and Bonney contributed to an ongoing misunderstanding of whaling custom still evident in current explanations of nineteenth-century practices. It was, however, the obscure structure and reasoning of Lowell's opinion that must bear most of the blame for the subsequent confusion as to the holding in *Swift*. Almost immediately, legal commentators incorrectly interpreted Lowell's ruling as measuring the timeliness of a first striker's claim by the point at which the eventual slayer cuts into a whale. In his 1881 *The Common Law*, Oliver Wendell Holmes, Jr. cited *Swift* for the proposition that American whalemen plying the Arctic Ocean award "a whale to the vessel whose iron first remains in it, provided claim be made before cutting in." Lost in Holmes' account was Lowell's holding that continuing pursuit – however that may be measured – was determinative of the rights of a ship that struck then lost a whale that was ultimately killed by another vessel. While cutting in might have been a reasonable and convenient measure of continued pursuit, whalemen do not appear to have employed this test.[20]

Subsequent scholars in the late nineteenth and early twentieth centuries largely followed Holmes in conflating the customs for pursuit of living whales with that for carcasses discovered adrift in the water. The influence and importance of Holmes as an interpreter of the common law may well have led other commentators to simply repeat the justice's assessment of *Swift* without further consideration of the case. In 1887, British barrister Whitley Stokes, for example, quoted Holmes – but not *Swift* – for the proposition that whalemen honored the claim of a first striker made before cutting in. Stokes' discussion

[19] Francis Allyn Olmstead, *Incidents of a Whaling Voyage. To Which are Added Observations on the Scenery, Manners and Customs, and Missionary Stations of the Sandwich and Society Islands, Accompanied by Numerous Lithographic Prints* (New York: D. Appleton and Co., 1841), 158–9.

[20] *Swift v. Gifford*, 23 F. Cas. 558. Oliver Wendell Holmes, Jr., *The Common Law* (Boston: Little, Brown, and Co., 1881), 212.

was also one of the earliest declarations that American judges such as Lowell followed industry custom in deciding whaling property disputes. As with most analyses of whaling cases, Stokes correctly noted that custom, not common law, was used to settle conflicts over contested whales. Stokes was, however, also typical in misconstruing the custom used and, most importantly, accepting the statement of that usage as set forth in the judge's opinion as an accurate assessment of how whalemen operated at sea. Recent scholars – while certainly more skeptical than their predecessors of the degree to which judicial opinions can be relied upon in understanding historic usages – have largely followed Holmes in accepting cutting in as the test for timeliness when two ships pursued a live whale. Robert Ellickson, whose interpretation of nineteenth-century American whaling custom has been uniformly accepted, offers the stipulation of counsel in *Swift* as evidence for the now venerable position that the cutting in test for timeliness of a claim was followed in the Sea of Okhotsk.[21]

III

The United States District Court in Massachusetts was not the only court in which American whalemen sought recourse for disputes that arose in the Pacific Ocean over possession of whales. Beginning in the 1840s, the growing community of English and American merchants in Hawaii started to bring commercial disputes before the local courts. The burgeoning western influence is reflected in the speed with which the Hawaiian legal system largely adopted American

[21] Whitley Stokes, *The Anglo-Indian Codes* (Oxford: Clarendon Press, 1887), 257–8. Robert C. Ellickson, *Order without Law: How Neighbors Settle Disputes* (Cambridge: Harvard University Press, 1991), 196–9. Other examples of reading *Swift* as adopting cutting in as a test for timeliness include: *The Lawyers Reports Annotated* 50 (Rochester: The Lawyers' Co-operative Publishing Co., 1903), 511; Henry Phillip Farnham, *The Law of Waters and Water Rights* (Rochester: The Lawyers' Co-operative Publishing Co., 1904), 1420; and Judah P. Benjamin, *A Treatise on the Sale of Personal Property; with References to the American Decisions and the French Code and Civil Law* (Boston: Houghton, Mifflin and Co., 1881), 194. Benjamin was the former Confederate Attorney General, Secretary of War, and Secretary of State turned English barrister. For Benjamin's legal career, see Eli N. Evans, *Judah P. Benjamin: The Jewish Confederate* (New York: The Free Press, 1988). For more recent misreadings of *Swift*, see, for example, Benjamin L. Fine, "Analysis of the Formation of Property Rights Underlying Tortious Interference with Contracts and Other Economic Relations," *University of Chicago Law Review* 50 (1983): 1116, 1131; and L. F. E. Goldie, "Mineral Rights and the General International Law Regime of the Deep Ocean Floor," *Brooklyn Journal of International Law* 2 (Fall 1975): 1,13. For another early commentator who cited and followed Holmes' interpretation of Swift, see Thomas Erskine Holland, *The Elements of Jurisprudence* (New York: Oxford University Press, 1917), 195. Some nineteenth-century commentators correctly recognized that *Swift* adopted a pursuit test. See, for example, William Wait, *A Treatise upon Some of the General Principles of the Law, Whether of a Legal, or of an Equitable Nature, Including their Relations and Application to Actions and Defenses in General, Whether in Courts of Common Law, or Courts of Equity; and Equally Adapted to Courts Governed by Codes* (Albany, New York: William Gould, Jr., & Publishers, 1885), 6:623.

commercial law during the middle decades of the nineteenth century. Therefore, when American whaling captain John Heppingstone brought an action in 1863 before the Supreme Court of Hawaii, he encountered an institution and a body of laws that would have been, in substantial part, familiar to any lawyer standing before the federal bench in Boston.[22]

Heppingstone's suit arose out of a dispute with John Mammen, the master of the German whaler *Oregon*, concerning a bowhead killed near Jonas Island in the Sea of Okhotsk. On May 29, 1863 boats belonging to the *Oregon* spotted a whale and gave chase. They succeeded in striking the bowhead and for the next two hours were dragged by the whale through floating ice. The *Oregon* was forced to cut its line when the whale disappeared beneath thick ice. When the whale unexpectedly rose again a few minutes later, one of the *Oregon's* boats again managed to strike and attach itself to the bowhead. That boat was stoved, but three other boats took up the chase and secured lines to the whale. Another boat was stoved and a second crew cut loose to pull their comrades out of the water. The harpoon of the *Oregon's* third boat came loose and the whale disappeared. Confident that the whale – spouting thick blood from the combination of hand and bomb lance strikes – was seriously injured, the *Oregon* decided to remain in the vicinity and renew the hunt at daybreak. To track the currents and hopefully signal the location of the whale they expected to expire during the night, a crew attached a waif pole to a block of floating ice. With daylight, the waif was recovered, but the whale had disappeared. The *Oregon* was left with little option but to sail in the direction which the bowhead likely made its surprising escape. Heavy fog which limited visibility to less than a mile lifted by about 9:00 AM and mate Antoine Costa went aloft with a spyglass to survey the water. After an hour, Costa finally spied a whale at a distance of about two to three miles with what appeared to be harpoons in its back. The whale, however, was not alone. Heppinstone's bark, the *Richmond*, was, at a mile and a half or so from the whale, also visible through the spyglass lowering its boats. Captain Mammen sent three boats to join the chase for what he assumed to be his bowhead. Costa was also ordered to inform the *Richmond* that the whale belonged to the *Oregon* and request that they desist in their attempts to take the animal.[23]

[22] For an introduction to the Hawaiian legal system in the nineteenth century, see Wendie Ellen Schneider, "Contentious Business: Merchants and the Creation of a Westernized Judiciary in Hawaii," *Yale Law Journal* 108 (April 1999): 1389–1424; Sally Engle Merry, *Colonizing Hawai'i: The Cultural Power of Law* (Princeton: Princeton University Press, 2000); and Jonathan Kay Kamakawiwo'ole Osorio, *Dismembering Lahui: A History of the Hawaiian Nation to 1887* (Honolulu: University of Hawai'I Press, 2002). For the development of nineteenth-century ideas in Hawaii about *ferae naturae* and property rights, see Wayne Tanaka, "Ho'ohana aku, Ho 'ola aku: First Steps to Averting the Tragedy of the Commons in Hawai'I's Nearshore Fisheries," *Asian-Pacific Law & Policy Journal* 10 (2008): 235–91.

[23] *Heppingstone v. Mammen*, 2 Hawaii 707 (Supreme Court of Hawaii, 1863). Testimony of Horace S. Allen, Antoine Costa, William Lipson, John Heppingstone, and John Mammen, Trial

Heppingstone indicated that although he was aware of harpoons in the back of the whale as his crews lowered, he had no reason to believe that they belonged to the *Oregon*. There were several other ships in the vicinity and the *Oregon* – within lowering range of the whale – initially showed no signs of joining the chase. The bowhead, in addition, did not – in the estimation of Heppingstone and his crew – appear to have been significantly slowed by the attached irons. Closer to the whale and with a head start of ten to twenty minutes in lowering, the boats of the *Richmond* reached the whale first. With the whale having sounded, the *Oregon's* third mate, Horace Allen, caught up with the boat carrying Captain Heppingstone and – indicating his claim to the animal – asked if the bowhead had several irons affixed to its back. Heppingstone acknowledged the presence of harpoons, but ordered his men to continue the hunt. When the whale surfaced a short time later, the boat commanded by the *Richmond's* third mate attached a harpoon and commenced lancing the animal. Gravely injured, the bowhead managed to find cover under nearby ice, forcing the *Richmond's* crew to cut the line. The whale came up for air in a patch of open water and was quickly dispatched by the *Richmond*.[24]

Negotiations over the whale between Heppingstone and the mates of the *Oregon* began even before the bowhead was killed. According to witnesses from the *Oregon*, Heppingstone initially stated that if the irons in the whale were from the *Oregon*, the animal belonged to them. Heppingstone denied making this statement, but the trial judge – Justice George Robertson of the Supreme Court of Hawaii – indicated in his opinion that the evidence showed that the captain of the *Richmond* had, indeed, said something to this effect. With several ships in the vicinity, Heppingstone perhaps felt that Costa was mistaken or was attempting to trick the *Richmond* into breaking off its pursuit. Once the whale was killed, Heppingstone suggested, as a compromise, that they "go halves." Costa insisted, however, that the matter be resolved by cutting the irons out of the whale. When the first iron proved to be unmarked, Heppinstone proclaimed "ah, you havent got that whale yet and you shant have that iron." The second iron bore the marks of a third ship. Heppingstone was not convinced by the explanation that the *Oregon* had acquired the irons from the other vessel and repeated his taunt that his rival had not demonstrated its right to the whale. Costa insisted, despite Heppingstone's reluctance, that a third

Transcript, *Heppingstone v. Mammen*, Admiralty #56, Hawaii State Archives, Honolulu, Hawaii (HSA). A hand lance has a wide blade attached to a wooden pole used to kill a whale. Bomb lances came into use after about 1850 and were shot at the whale from a shoulder gun. Having penetrated the whale, a delayed charge exploded. While an obvious improvement in that they could be fired with accuracy from up to about sixty feet from the target, bomb lances were extremely dangerous to use. John R. Bockstoce, *Whales, Ice, and Men: The History of Whaling in the Western Arctic* (Seattle: University of Washington Press, 1986), 62–4.

[24] *Heppingstone v. Mammen*, 2 Hawaii 707. Testimony of John Heppingstone, Horace S. Allen, Jonathan Rogers, and Robert Chapel. Trial Transcript, *Heppingstone v. Mammen*, Admiralty #56, HSA.

iron be removed. Heppingstone cut out another iron, glanced at the marking, covered it with his hand, and proposed three times "what do you say halves." Horace Allen, the *Oregon's* third mate, looked at the iron and announced that it belonged to his ship. Heppingstone proclaimed that he did not know if by law he was entitled to half the whale, but that for all his trouble he should be granted a share. When Costa indicated that it was not for him to decide, Heppingstone prepared to board the *Oregon*, calling Captain Mammen "a damned Dutchman" and declaring that most American ships would willingly share the whale. Heppingstone's parting comment was that "if the captain is not man enough to give me half of it I would just as have tell him to kiss my arse as not." Negotiations on board the *Oregon* ended with Mammen refusing Heppingstone's request and the latter's declaration that the matter would be revisited when they returned to Hawaii.[25]

The bargaining continued in Honolulu on the day after the *Oregon* returned to port. On November 14, 1863 Mammen and Heppingstone met in the store of Captain P. S. Wilcox. It appears from Mammen's testimony that Wilcox, who testified at trial as an expert witness for the *Richmond*, was attempting to mediate the dispute on Heppingstone's behalf. Wilcox asked Mammen if there was some sort of settlement that could be reached that would save the expenses of litigation. Mammen was adamant that he would not be a part of any arrangement that would divide possession of the whale. Declaring that he was through dealing with Mammen, Wilcox asked the *Oregon's* captain for the name of the ship's agent. Mammen proclaimed that if the agent, Mr. Thomas, "gave a drop of oil from that whale," he would resign his position. Whether Wilcox pursued the matter with Thomas is unknown.[26]

Heppingstone testified at trial that he knew of no custom governing a whale that is struck by one ship and then killed by another vessel. He did indicate, however, that if he saw that the first striker remained in pursuit of a whale bearing three of its irons he would not join the hunt. Heppingstone suggested that the *Oregon* – rather than making a verbal claim – should have attempted to kill the disputed bowhead which, prior to its slaying, was free for the taking. If Heppingstone seemed to advocate iron holds the whale as advisory and, in any event, applicable only when the first striker was clearly in pursuit of a whale struck several times, Captain Wilcox presented a strict version of fast-fish, loose-fish. There exists, Wilcox explained, no rule to prevent the party that kills a whale from claiming it. Wilcox indicated that it was common to kill and unremarkable to claim a bowhead wounded the previous day by another ship. Third mate Jonathan Rogers of the *Richmond* opined that pursuit of a

[25] *Heppingstone v. Mammen*, 2 Hawaii 707. Testimony of Horace Allen, John Mammen, Calenzo Pedro, and John Heppingstone. Trial Transcript, *Heppingstone v. Mammen*, Admiralty #56, HSA.

[26] Testimony of P. S. Wilcox, John Mammen, and John Heppingstone. Trial Transcript, *Heppingstone v. Mammen*, Admiralty #56, HSA and *Friend* (Honolulu), December 1, 1863.

wounded whale with attached irons holds the animal, but that in the absence of such pursuit the prize was commonly shared. When Heppingstone boarded the *Oregon* to speak with Captain Mammen, Rogers expected that the whale would be divided as "the practice is to give half without any grumbling where I have been." Captain Thomas Williams – an expert witness called by the *Richmond* – provided a particularly convoluted explanation of the applicable customs. He declared that a ship that strikes a whale, but claims it the following day is only entitled to the return of its craft from the slayer. While a first striker's claim to a whale struck the previous day is more compelling if he warned the second vessel of his claim before the latter affixed its first harpoon, the slayer is still entitled to the whale if he was the first to lower his boats. The outcome would be different, in Williams' estimation, if the two ships lowered about the same time. In that scenario, the whale should be divided. The *Oregon's* sole expert witness, Captain F. W. Wilbur, stated that there is no custom or usage that prevents the first striker from claiming a whale bearing the ship's irons. Wilbur revealed how his ship once willingly relinquished a whale it had killed when another vessel made a claim based on its affixed irons. "It is," Wilbur intoned, "customary with me to give up any thing that belongs to another man." Wilbur concluded his testimony with the apparent caveat that he had stated the custom among German masters and that "I don't know what it may be elsewhere." If German captains did, indeed, uniformly follow a single standard, their American brethren showed no immediate sign of following their example.[27]

Two additional expert witnesses called by Heppingstone – Captains J. Sowle and John Dexter – seem to call into question the existence of what could even be considered customs in the Pacific concerning the pursuit of live whales. Captain Sowle, noting that he would take a whale struck the day before by a ship whose harpoons remained affixed, observed that "by the custom I can not say to whom it belongs." Sowle elaborated that he knew of no custom that would favor a first striker even if he provided notice of his claim prior to the whale's death. As with Heppingstone's explanation of custom, Sowle depicted customs as being, if not personal, limited in scope. He revealed when asked about a particular usage: "I know of nothing of the custom any more than hearsay. I can not say what other captains minds might be." Captain Dexter shared his colleague's uncertainty in replying hesitantly to a similar query, "that is what I understand to be the custom."[28]

If these witnesses expressed divergent views and, perhaps, even doubt as to the existence of custom concerning living whales, they were unanimous in describing practices regarding dead cetaceans. Heppingstone, Wilcox, Rogers,

[27] Testimony of John Heppingstone, F. W. Wilbur, Thomas Williams, Jonathan Rogers, and P. S. Wilcox, Trial Transcript, *Heppingstone v. Mammen*, Admiralty #56, HSA.

[28] Testimony of J. Sowle and John Dexter, Trial Transcript, *Heppingstone v. Mammen*, Admiralty #56, HSA.

and Williams stated with great certainty that a discovered whale carcass is split between the finder and a claimant whose irons remain affixed. Wilcox explained that "there is a distinction between a dead and a live whale on the whaling grounds. If I picked up a dead whale I would have held on to all I had cut in and let the balance below the plankshear go to the person who claimed it. If I had cut in the whole whale I could keep the whole whale." This was precisely the standard which the attorneys in *Swift* applied to a living whale. While the stipulation to this standard in *Swift* was, as we have seen, made by attorneys for reasons that likely had more to do with legal strategy than concern for actual practices, their source was the expert testimony of an experienced Sea of Okhotsk master. The question is thus raised whether Captain Abraham W. Pierce was mistaken, deceptive on behalf of Gifford, or testified accurately as to his own experience? Deposed almost seven years after the trial in *Heppingstone*, Pierce may have stated his understanding of an evolving practice that was moving toward the use of the point at which a ship claimed a whale as the test of timeliness pursuant to the holding power of an iron in a living whale. Perhaps the unanimity expressed by the four whalemen in the proceedings in the Supreme Court of Hawaii was a reflection of an earlier custom that was – even as they testified – changing. That the custom was moving in this direction in the 1860s seems unlikely as no supporting evidence other than that presented in *Swift* has been discovered. Assuming that Pierce testified in good faith, he was either mistaken or explaining a highly localized custom the observance of which was limited to a small section of the North Pacific.[29]

Upon the conclusion of testimony, counsel for Heppingstone argued that the matter should be resolved by application of the common law doctrine of *ferae naturae*. Two cases from the Supreme Court of Judicature of New York were cited for the proposition that property in a wild animal is acquired by occupancy. In *Buster v. Newkirk*, a hunter was said to occupy – or possess – a deer if he deprived it of its "natural liberty" by bringing the animal within his "power and control." The facts in *Buster* were, as Heppingstone's counsel likely argued, particularly on point with the case at bar. The deer was wounded but managed to run off with Newkirk, the hunter, in pursuit. Although darkness forced Newkirk to temporarily abandon the chase, his dogs continued on the trail of the bleeding deer. During the night, a third party fired and struck the deer which, with one of Newkirk's persistent hounds closing in, ran onto the property of Buster. The deer's six mile dash for safety ended when Buster slit its throat. With the morning light, Newkirk resumed pursuit which ended at Buster's home with a rebuffed claim of the hide. The court ruled that the abandoned pursuit and the ability of the deer to run six miles after being

[29] Testimony of John Heppingstone, Thomas Williams, Jonathan Rogers, and P. S. Wilcox, Trial Transcript, *Heppingstone v. Mammen*, Admiralty #56, HSA.

wounded demonstrated that Newkirk had not deprived it of its natural liberty. The second New York case cited by Heppingstone, *Gillet v. Mason*, held that the mere discovery and marking of a tree containing a swarm of bees does not constitute possession. It is the hiving of the bees which reduces the insects to private property.[30]

Mammen countered that as his voluntary abandonment of the chase was necessitated by the exigency of saving his crew, the court should deem his pursuit of the whale continuous. The *Oregon*, having inflicted a serious injury the previous day, was able to overtake the whale prior to the first strike of the *Richmond* and would have certainly killed its prey were it not for the interference of Heppingstone's vessel. The warning provided to Heppingstone that the whale belonged to the *Oregon*, in addition, placed the *Richmond* on notice that any actions taken in pursuit of the whale would be on Mammen's behalf. The remainder of Mammen's arguments asserted that Heppingstone's claim on the whale should be estopped by his statements and actions at sea. Mammen argued that prior to removing the harpoons Heppingstone conceded that the whale belonged to the *Oregon* should the irons bear its mark. Heppingstone's decision to allow the *Oregon* to render the bowhead was, in Mammen's estimation, a clear indication that he recognized that the whale belonged to the *Oregon*. Mammen further contended that any recovery by Heppingstone be limited to the one half of the whale's value he claimed in the heat of the dispute.[31]

Presented with a common law governing wild animals that was far from settled on the crucial point of what constituted sufficient occupancy for ownership and the contradictory statements of whaling custom, Justice Robertson was left without clear guidance as to the applicable law. Robertson commenced his discussion of the law by observing, without explanation, that English common law applied only to events transpiring in the Kingdom of Hawaii. Disputes at sea were, on the other hand, to be decided by the applicable dictates of maritime law. In the absence of any provisions in maritime law governing this situation, Robertson relied on the guidance of what he termed "equity and natural right." While Robertson never identified any particular provisions of either equity or natural right in his opinion, he used whaling custom analyzed through the lens of fairness to render a verdict. Robertson reduced the welter of conflicting testimony as to whaling custom to a pair of basic principles. A dead whale belongs entirely to a finder who can cut in prior to a verifiable claim made by the slayer. A wounded whale belongs to its slayer if that ship lowers and kills the animal before the first striker joins in the pursuit. The problem in the case at bar was, as Robertson explained, that the *Richmond* was placed on notice prior to striking the bowhead that the *Oregon* claimed the whale based

30 Trial Transcript, *Heppingstone v. Mammen*, Admiralty #56, HSA. *Buster*, 20 Johns. 75. *Gillet v. Mason*, 7 Johns. 16 (1810).

31 Trial Transcript, *Heppingstone v. Mammen*, Admiralty #56, HSA.

upon an identification of the affixed harpoons. The *Oregon* was, in addition, also in position at the time of its claim to take the whale. Although Robertson acknowledged that testimony was presented that even under these complicating circumstances the custom followed by American whalemen would award the whale to the *Richmond*, the justice explained that he found such a usage unreasonable. At this point, Robertson moved from custom and the common law as modified by usage and turned to fairness.[32]

To explain why he found a custom that would give the entire whale to the *Richmond* unfair, Robertson posed a hypothetical. Suppose, Robertson queried, that the *Oregon* had relinquished the bowhead at sea to its competitor and thus Mammen – rather than Heppingstone – had brought this action. Robertson asked if, in this scenario, it would be unfair not to reward the *Oregon* for the danger encountered in disabling the whale and its persistence in renewing the chase and placing itself ultimately in a position to strike a fatal blow. Having answered this question in the affirmative, Robertson turned back to the matter at hand and asked if "in equity and good conscience" the *Oregon* should be allowed to retain the entire whale. The *Oregon* had, after all, been forced to cut its lines and the whale – though wounded – was free when killed by the *Richmond*. Had Heppingstone's men not intervened, the bowhead would likely have evaded capture. The odds, Robertson offered in regards to the whale's prospects for escaping the *Oregon*, "were rather in its favor." Satisfied that the parties had each stated a compelling right to the bowhead, Robertson concluded that "under all the circumstances of the case, the whale may fairly be considered the joint prize of both ships."[33]

IV

Evidence of the negotiations in whaling disputes at sea and on land has survived primarily from matters that were litigated. There is no way to know to what degree such bargaining was common. The testimony in *Heppingstone* certainly suggests that it was. Various crew members and expert witnesses provided their understanding of the prevailing customs of whaling. They expressed such uncertainty as to the governing norms and provided such a welter of conflicting standards that it seems that ambiguity was the sole constant. Beyond a general statement that in certain circumstances an affixed iron in an injured whale

[32] Opinion, Trial Transcript, *Heppingstone v. Mammen*, Admiralty #56, HSA. *Heppingstone v. Mammen*, 2 Hawaii 707, 711. There is some evidence that the lowering of the first boat was, on occasion, used to determine which ship was entitled to a whale. John Bockstoce believes that whalemen in the Western Arctic agreed to honor the claim of the ship which was able to put the first boat into the water. Bockstoce, *Whales, Ice, and Men*, 61. Ellickson has rightly questioned Bockstoce's reliance on a single childhood memory recorded many years later. Ellickson, *Order without Law*, 196.

[33] Opinion, Trial Transcript, *Heppingstone v. Mammen*, Admiralty #56, HSA. *Heppingstone v. Mammen*, 2 Hawaii 707, 713, 714.

might preserve some rights in the first striker, it is difficult to draw anything resembling a rule. In an atmosphere where whalemen appeared, at times, to be making things up as they went along, it is hard to imagine that such negotiations were not an accepted practice.

The evidence in *Heppingstone* and the other cases also suggest that these negotiations were not, strictly speaking, exclusively about applicable customs, norms, or usages. When Heppingstone stated at sea that he believed that he was entitled to a share of the whale for his trouble, he acknowledged that his claim was not based upon law or custom. Similarly, in court, Heppingstone indicated that he was not aware of any custom governing the situation. The belief of *Richmond* mate Jonathan Rogers that the *Oregon* would readily hand over half the whale without complaint was also based on fairness. Heppingstone's sense that effort should be rewarded – even in the absence of a compelling claim based on industry custom – was shared by Silvey and Captain Howland in *Swift*. Silvey and Howland argued that the *Hercules* should be granted half the whale as compensation for its decision to forego another whale and kill a bowhead they had no reason to believe belonged to another vessel. As with the arguments of Heppingstone and Rogers, principles of fairness – not custom – animated the claims of Silvey and Howland. The sense that each of these whalemen was referring to a code of conduct or way of measuring rights to a disputed cetacean that was as much personal as it was communal was captured in the testimony of Captain Wilbur. "It is," Wilbur testified in *Heppingstone*, "customary with me to give up any thing that belongs to another man."[34]

The terms custom, norm, and usage are often used somewhat interchangeably throughout the voluminous literature discussing socially approved behavior that is created and enforced outside the dictates and formal institutions of the state. Legal scholars, however, generally use the word norm when they attempt to rigorously define and examine the concept of rules of conduct that originate with participants in the involved business or community. Use of the term custom, perhaps, invites confusion with the process by which community practices of long standing come in common law to enjoy recognition as law. While the offered definitions of a norm varies, the salient point in all characterizations of the concept is that a norm is not recognized as binding positive law. As legal scholar Lawrence Mitchell has observed, a norm is an "ought statement." "[N]orms," Mitchell explains, "tell us what we should do under a given set of circumstances and are therefore obligatory upon those who wish to participate in the society which is at least partly constituted by such norms." Norms, of course, are not the only force that molds, directs, and coerces human behavior. Robert Ellickson has provided a useful discussion of other sources of

[34] Testimony of John Heppingstone, Jonathan Rogers, and F. W. Wilbur Trial Transcript, *Heppingstone v. Mammen*, Admiralty #56, HSA. Depositions of George Silvey and Robert D. Gifford, Case File, *Swift v. Gifford*, National Archives, Northeast Region.

behavioral rules and means of enforcement which he terms controllers. A government sets forth laws and has the authority and means to insure compliance. Organizations such as the Roman Catholic Church are also the source of rules governing actions and imposing sanctions on those who voluntarily place themselves under its authority. Parties who enter a contract have similarly chosen to act or refrain from acting in an agreed upon manner. Some people operate under a set of personal ethics which are self-enforced by that individual's desire to act in a way he or she deems proper. A self-imposed code may be barely distinguishable from prevailing norms in the surrounding society or it may be highly idiosyncratic.[35]

What distinguishes norms in the scholarly literature from personal ethics is that members of the community understand what behavior a norm requires. Absent such an understanding, a norm will not – at least in theory – function efficiently. The enforcement mechanism of a norm will break down if community members cannot agree as to the sort of behavior to punish. In the case of nineteenth-century whaling in the Sea of Okhotsk, the rules concerning dead whales discovered adrift were clearly norms. Whalemen were able to provide consistent statements as to who was entitled to a dead cetacean. The norm even provided a rather precise means for determining the percentage of the whale to be awarded based on the progress of the cutting in process at the time the claim was made. A captain who violated the norm for dead whales did so with the clear knowledge that he risked his good standing in the community of whalemen. The provisions governing live whales bearing the irons of a competing vessel were not, however, so easily expressed. As the testimony in *Swift* and *Heppingstone* makes clear, there were almost as many understandings of the governing rules as there were witnesses. While some of the explanations sound very much like statements of personal ethics, there were, nevertheless, common themes that run through the testimony. A first striker that maintains pursuit

[35] Ellickson, *Order without Law*, 126–32. Lawrence E. Mitchell, "Understanding Norms," *University of Toronto Law Journal* 49 (1999): 177–248, 181–2. For a taste of the huge literature concerning norms, see Steven A. Hetcher, *Norms in a Wired World* (New York: Cambridge University Press, 2004); Paul G. Mahoney and Chris W. Sanchirico, "Competing Norms and Social Evolution: Is the Fittest Norm Efficient?" *University of Pennsylvania Law Review* 149 (2001): 2027–62; Jonathan R. Macey, "Public and Private Ordering and the Production of Legitimate and Illegitimate Legal Rules," *Cornell Law Review* 82 (1997): 1123–49; Richard H. McAdams, "Accounting for Norms," *Wisconsin Law Review* 1997, no. 3 (1997): 625–37; C. A. Harwell Wells, "The End of the Affair? Anti-Dueling Laws and Social Norms in Antebellum America," *Vanderbilt Law Review* 54 (2001): 1805–47; Richard A. Epstein, "*International News Service v. Associated Press:* Custom and Law as Sources of Property Rights in News," *Virginia Law Review* 78 (1992): 85–128; Dan M. Kahan, "Signaling or Reciprocating? A Response to Eric Posner's *Law and Social Norms*," *University of Richmond Law Review* 36 (2002): 367–85; Henry E. Smith, "Community and Custom in Property," *Theoretical Inquiries in Law* 10 (2009): 5–41; and Gregory M. Duhl, "Property and Custom: Allocating Space in Public Places," *Temple Law Review* 79 (2006): 199–248.

was, for instance, deserving of some consideration over an intervening vessel. Iron, in other words, sort of held the whale.[36]

If the rules for taking wounded whales were not sufficiently clear to satisfy the definition of a norm, what were they? It is perhaps most useful to view the practices in the Sea of Okhotsk as a sort of hybrid between personal ethics and norms. Whalemen such as Heppingstone, Rogers, Silvey, and Howland each saw the code of conduct governing their behavior as both personal and communal. In the absence of a prevailing norm, whalemen in the Sea of Okhotsk were left with the guidance of their own personal ethics as refracted through the prism of old norms and the exigencies of the communal business of chasing whales in the company of a close-knit brotherhood. This is precisely what Captain Sowle had in mind when he testified in *Heppingstone* that his understanding of custom was based on hearsay rather than firsthand experience of the usage in practice. Although Sowle could not state "what other captains minds might be," he, nevertheless, had a notion of how other masters thought about and resolved disputes over wounded whales. That the rules in the Sea of Okhotsk were part personal ethic and part norm should not be surprising. Even in the Greenland whalefishery of the early nineteenth century the generally understood norm of fast-fish, loose-fish was tempered by what Scoresby called the "laws of honor," which were a mix of Christianity's Golden Rule and the captain's individual notion of proper behavior.[37]

Yet, if whaling practices in the Sea of Okhotsk were confusing and not well understood by the men who hunted cetaceans, why were they so remarkably successful in keeping disputes out of court? The answer is likely multifaceted. The transaction costs of litigation were exceptionally high both in terms of the price of bringing or defending a cause of action and in the amount of time required before a verdict would be rendered. As was previously suggested, those American disputes that were litigated often had an element that made those conflicts a matter of personal principle. In *Taber*, for example, the attempt to fix the result of arbitration likely motivated the appeal more than strictly monetary calculations. Compromise – given the high transaction costs and the unwillingness to engage in violent forms of self-help – also made good economic and business sense. Whalemen had, in addition, a long history of sharing whales. When Heppingstone suggested to the mate of the *Oregon* that

[36] Testimony of John Heppingstone, F. W. Wilbur, Thomas Williams, Jonathan Rogers, J. Sowle, John Dexter, and P. S. Wilcox, Trial Transcript, *Heppingstone v. Mammen*, Admiralty #56, HSA. Depositions of George Silvey and Robert D. Gifford, Case File, *Swift v. Gifford*, National Archives, Northeast Region.

[37] Testimony of John Heppingstone, F. W. Wilbur, Thomas Williams, Jonathan Rogers, J. Sowle, John Dexter, and P. S. Wilcox, Trial Transcript, *Heppingstone v. Mammen*, Admiralty #56, HSA. Depositions of George Silvey and Robert D. Gifford, Case File, *Swift v. Gifford*, National Archives, Northeast Region. C. Ian Jackson, ed., *The Arctic Whaling Journals of William Scoresby the Younger: Volume I, The Voyages of 1811, 1812 and 1813* (London: The Hakluyt Society, 2003), 119.

they "go halves" he echoed the words of Scoresby in his 1812 confrontation with his father and drew on a tradition that included seventeenth-century regulations for splitting drift whales, the late eighteenth-century Galapagos custom of dividing whales between first striker and slayer, eighteenth-century New England usage of awarding a fractional share to a competitor who answered the call for assistance, and the widespread practice of mateship. Ellickson is undoubtedly correct in insisting that close-knit communities are often capable of managing their affairs without recourse to courts and other formal manifestations of the law. What he overestimates is the necessity for well understood norms in settling disputes. Whalemen in the Sea of Okhotsk proved adept at resolving controversies on a common sense, ad hoc basis without universal norms. The close-knit nature of their community, the intensely communal nature of their competition, and the economic pressure to settle disputes allowed Okhotsk whalemen to resolve contests without the aid of well settled norms.[38]

[38] Ellickson, *Order without Law*, 191–206. For the seminal use of the concept of transaction costs, see Ronald Coase, "The Problem of Social Cost," *Journal of Law and Economics* 3 (1960): 1–44. See also, Pierre Schlag, "The Problem of Transaction Costs," *Southern California Law Review* 62 (1989): 1661–1700.

7

The environment of litigation

The population of bowheads in the Sea of Okhotsk was relatively small and was hunted by whalers for only about twenty years. Why all of the five reported American cases in the nineteenth century involving possession of a whale should have originated from this particular fishery has long remained a mystery. Robert Ellickson has suggested that the litigation was caused by the transition from pursuit of the aggressive and combative sperm whale to the slower and relatively docile bowhead. Iron holds the whale was ideal for a sperm whale fishery, Ellickson postulates, but was inefficient for the taking of bowheads. If this transition, however, disrupted the dispute resolution techniques of American whalers in the Sea of Okhotsk, why did the bowhead fishery of the Western Arctic in the same period not also send disputes to court? The Western Arctic fishery was, after all, conducted by many of the same men who hunted the Sea of Okhotsk. The answer to the question of why whalemen in the Sea of Okhotsk resorted to litigation after 1850 can be found by looking at bowhead biology and the currents, zooplankton, and ice formation in the waters west of Kamchatka and north of the Sea of Japan. It was the confluence of these environmental factors in the Sea of Okhotsk in conjunction with the economic exigencies felt by all whaling captains that explains the curious litigiousness of the Sea of Okhotsk.

I

Why did the successful system for extralegal dispute resolution break down after 1850 resulting in the only whaling property contests litigated in nineteenth-century American courts? Robert Ellickson provides two possible explanations for the sudden spate of litigated cases all stemming from conflicts in the bowhead fishery of the Sea of Okhotsk between 1852 and 1867. The first involves the transition during the first half of the nineteenth from the pursuit of sperm whales off the west coast of South America and along the equator to the bowhead and right whale fisheries in the North Pacific and Arctic Ocean. Whalemen – the theory goes – adopted the custom of iron holds the whale to maximize the take of the ornery and swift sperm whales. The aggressive sperm whales of the Southern fishery rendered the fast-fish, loose-fish requirement of continuous connection impractical and too dangerous. Slow and relatively docile, bowheads and right whales presented whalemen with different challenges that were, in turn, ill suited to the custom of iron holds the whale. Whalemen, in a period of transition from sperm whale to bowhead hunting practices, were suddenly unable in the Sea of Okhotsk to settle disputes on their own.

There is much to like about this theory. It offers historians, anxious to introduce whale ecology and agency, an alternative to the prevalent stories of whaling as an adventure yarn or as a declension narrative of human depredation. It is, however, inaccurate. Hunting techniques were remarkably consistent from the seventeenth century to the last quarter of the nineteenth century. There were, of course, changes over time in how whalemen pursued their prey, but the custom of iron holds the whale which Ellickson views as a creation of sperm whalers was already practiced by Americans hunting right whales in the North Atlantic in the late eighteenth century. When American and British ships – largely under the command of captains from Nantucket – turned south in the late eighteenth century to chase sperm whales they carried with them their preference for iron holds the whale. Iron holds the whale was, in fact, well suited for any species of cetacean chased by men in boats with ropes, harpoons, and lances.[1]

Ellickson's second explanation is that by the 1850s the close-knit community of American whalemen was losing its cohesion. With whaling in decline, captains and ship owners began to defect from community customs in the belief that such practices were no longer in their best interests. Realizing that the industry was in serious decline whalemen decided – in the classic tragedy of the commons scenario or in accordance with Prisoner's Dilemma – to exploit a dying resource as quickly as possible without concern for the long-term health of the trade or their personal standing in the community. There is, however, no real evidence of a general loss of cohesion among the worldwide community of whalemen. Bad times certainly lay ahead, but the whaling industry remained generally profitable in the years prior to the Civil War. If the whaling community was losing its cohesion by the 1850s, it should have been felt throughout the industry. The whalemen who hunted the Sea of Okhotsk wintered in Hawaii with a similar number of captains who pursued bowheads in the

[1] Robert C. Ellickson, *Order without Law: How Neighbors Settle Disputes* (Cambridge: Harvard University Press, 1991), 191–206. For a discussion of the hazards of hunting bowheads in icy waters, see William Scoresby, *An Account of the Arctic Regions with a History and Description of the Northern Whale-Fishery* (Edinburgh: Archibald Constable and Co., 1820), 2:346–68.

Western Arctic. There appears to be no reason why these members of the same whaling community should have been immune from the desire to defect from community norms that allegedly befell their colleagues in the Sea of Okhotsk.[2]

II

The initial seasons hunting bowheads in the Sea of Okhotsk and the Western Arctic were, by all measures, exceptionally successful. In 1848, when Captain Thomas Roys of the *Superior* returned to Hawaii after conducting the first Anglo-American whaling in and above the Bering Strait with a full hold and tales of giant bowheads yielding in excess of 300 barrels of oil, his colleagues took notice. During the following season, approximately one-third of the North Pacific fleet of over 150 vessels was drawn deep into the Bering Sea on the evidence provided by a single cruise. The promise of a new fishery inhabited by whales of almost mythical size was potent enough to lure captains and crews into unknown waters that many believed posed potential dangers beyond that which even nineteenth-century whalemen were generally willing to accept. The risk paid off handsomely. The average ship returned to Hawaii with over 1300 barrels of oil. After a second season of success enjoyed by over 130 ships, whalemen appeared giddy as if some sort of promised land had been reached. The unusually fine weather and ice conditions did not last and the 1851 season was dismal. The failure of the whales to appear in their previous haunts in the Gulf of Anadyr and off Cape Navarin, forced the fleet of 170 or so vessels farther north. Seven ships were lost and the fleet returned home with a fraction of its previous hauls. In 1852, reflecting the hope born of earlier success and perhaps the lack of a more promising alternative, whalemen returned to the area of the Bering Strait in even greater numbers. When the meager harvest of 1852 was matched the following season, it appeared that the vision of a productive new fishery in the Bering Sea and latitudes to the north was more chimera than reality. Poor yields in 1854 and 1855 reduced the number of ships willing to try their luck in the Bering Sea to a mere handful in 1856.[3]

[2] Ellickson, *Order without Law*, 191–206. *Whalemen's Shipping List and Merchant's Transcript* (New Bedford, MA), February 1, 1853. Garrett Hardin, "The Tragedy of the Commons," *Science* 162 (December 13, 1968): 1243–8. For a discussion of Prisoner's Dilemma, see Ellickson, *Order without Law*, 159–62. The profitability of American whaling is dissected in Lance E. Davis, Robert E. Gallman, and Karin Gleiter, *In Pursuit of Leviathan: Technology, Institutions, Productivity, and Profits in American Whaling, 1816–1906* (Chicago: University of Chicago Press, 1997), 423–58.

[3] John R. Bockstoce, Daniel B. Botkin, Alex Philp, Brian W. Collins, and John C. George, "The Geographic Distribution of Bowhead Whales, Balaena Mysticetus, in the Bering, Chukchi, and Beaufort Seas: Evidence from Whaleship Records, 1849–1914, *Marine Fisheries Review* 67, issue 3 (2005): 1–43. John R. Bockstoce, *Whales, Ice, and Men: The History of Whaling in the Western Arctic* (Seattle: University of Washington Press, 1986), 21–9, 93–102; George W. Schuster, "Productivity and the Decline of American Sperm Whaling," *Environmental Affairs* 2 (1972–1973):

Despite the sudden bloom and rapid demise of whaling in the Bering Sea and Western Arctic by the middle of the 1850s, the fishery reemerged in the 1860s and – in the words of whaling historian John Bockstoce – "prolonged the life of the American whaling industry" into the first years of the twentieth century. With an estimated 1848 population of 15 to 25 thousand bowheads – based on the middle range of many often disparate studies – the Bering-Chukchi-Beaufort or Western Arctic stock was clearly not destroyed by a few successful seasons. There are several possible explanations for the grim harvests of the mid-1850s. John Bockstoce and his several co-authors suggest that the bowheads in this region – unfamiliar with the ways of intensive commercial hunting – required a bit of time to learn the warning signs of an imminent strike and adopt avoidance techniques such as evasive rolls. Whalemen in all fisheries always insisted that familiarity with humans taught whales how to avoid predators. The keen hearing of bowheads and their ability to communicate with one another over many miles supports what eighteenth- and nineteenth-century whalemen suspected. Humans also needed to adapt to the particularities of these new grounds. The natural tendency of whalemen was to return to the same areas where they had enjoyed previous success. When the whales could not be easily found, whalemen – unfamiliar with the vast range of the new fishery – did not initially know where to look. With experience and a willingness to sail deeper into the Beaufort Sea, whalemen learned where the whales had gone in the lean early years of the fishery.[4]

345–57, 350–1; Eric Hilt, "Investment and Diversification in the American Whaling Industry," *The Journal of Economic History* 67, issue 2 (June 2007): 292–314; and Davis, Gallman, and Gleiter, *In Pursuit of Leviathan*, 69, 376.

[4] Bockstoce, *Whales, Ice, and Men*, 93–102; Bockstoce, Botkin, Philp, Collins, and George, "The Geographic Distribution of Bowhead Whales," 1–12; and Douglas A. Woodby and Daniel B. Botkin, "Stock Sizes Prior to Commercial Whaling," in *The Bowhead Whale*, ed. John J. Burns, J. Jerome Montague, and Cleveland J. Cowles (Lawrence, KS: The Society for Marine Mammalogy, 1993), 387–407. For a general discussion of how whales use acoustic cues to identify and locate potential dangers and predators, see Brian K. Branstetter and Eduardo Mercado, III, "Sound Localization by Cetaceans," *International Journal of Comparative Psychology* 19 (2006): 26–61. The debate whether whales are capable of cultural transmission and evolution is far from settled, but the possibility that whales can learn and teach survival skills might explain some of what whalemen observed in the early years of a fishery. Luke Rendell and Hal Whitehead, "Culture in Whales and Dolphins," *Behavioral and Brain Sciences* 24 (2001): 309–24. For an open peer commentary on this article joined by many scholars and a response by Rendell and Whitehead, see *Behavioral and Brain Sciences* 24 (2001): 324–82. The possibility that the Bering-Chukchi-Beaufort stock of bowheads has had subgroups suggests that the first whalemen in the Western Arctic decimated a distinct population in the southern Bering Sea during the opening years of the fishery. Bockstoce, Botkin, Philp, Collins, and George, "The Geographic Distribution of Bowhead Whales," 1–12; and R. G. LeDuc, K. K. Martien, P. A. Morin, N. Hedrick, K. M. Robertson, B. L. Taylor, N. S. Mugue, R. G. Borodin, D. A. Zelenina, and J. C. George, "Mitochondrial Genetic Variation in Bowhead Whales in the Western Arctic," accessed September 25, 2015, https://archive.iwc.int/pages/preview.php?ref=267&alternative=1445&k=&search=!collection15&offset=0&order_by=relevance&sort=DESC&archive=0&.

While the Bering-Chukchi-Beaufort stock of bowheads in the Western Arctic fishery was subject to the same habitat and reproductive limitations as their mates in the Sea of Okhotsk, the former enjoyed a number of advantages that preserved – at least for a few decades – the stock. The area in which they roamed was immense. Bowheads were taken in this fishery from a point at about 57 degrees north and 165 degrees west, up along the Siberian coast, and through the Bering Strait. Beyond the Bering Strait, whalemen killed bowheads in the Chukchi Sea as far north and west as Wrangel Island. The fleet also ranged successfully to the east along the southern coast of the Beaufort Sea and into the Amundson Gulf. With numerous haunts known to whalemen, others to be discovered only after further exploration, and a sheltering northern edge of ice, the bowheads of the Western Arctic retained sufficient numbers to remain the subject of American commercial whaling until the second decade of the twentieth century. The possibility that the Bering-Chukchi-Beaufort stock feed in the late Autumn and Winter might also have allowed the whales – free of the biological necessity of spending the summer hunting season in only water with the highest biomass of zooplankton – to further evade their predators. The limited number of bowheads in the Sea of Okhotsk enjoyed none of these advantages and were severely depleted by 1870.[5]

The conditions and extent of the Western Arctic grounds worked to preserve the stock of bowheads and, thereby, fostered – at least in relative terms – a sustainable fishery. If whales could escape hunters, whalemen also had more options should the ice and wind render a particular gulf too dangerous. Whaling in the Bering Sea and beyond was not, as it was often in the Sea of Okhotsk, a matter of feast or famine. A captain in the Sea of Okhotsk who experienced difficult conditions around the Shantar Islands had few options. He could sail to the North-east Gulf or perhaps head south to hunt right whales in more temperate waters. Most captains, instead, chose to persevere in the vicinity of the Shantar Islands hoping for a change in the weather and the opportunity to renew the chase. Captains in the Western Arctic could simply move to another portion of the fishery enjoying more favorable conditions. There were, of course, years in the Western Arctic that were either more or less favorable for hunting, but once the ships returned after 1857 the fishery yielded – by whaling standards – remarkably consistent results until the early 1870s. Bockstoce and others have calculated the number of days it took on a yearly average to take a whale in the Western Arctic. Between 1858 and 1870,

[5] Bockstoce, Botkin, Philp, Collins, and George, "The Geographic Distribution of Bowhead Whales," 1–43. For the feeding habits of the Bering-Chukchi-Beaufort stock, see Donald M. Schell and Susan M. Saupe, "Feeding and Growth as Indicated by Stable Isotopes," in *The Bowhead Whale*, ed. John J. Burns, J. Jerome Montague, and Cleveland J. Cowles (Lawrence, KS: The Society for Marine Mammalogy, 1993), 491–509; and Donald M. Schell, "Declining Carrying Capacity in the Bering Sea: Isotopic Evidence from Whale Baleen," *Liminology and Oceanography* 45, no. 2 (March 2000): 459–62.

the yearly average in eight of the thirteen seasons fell in a tight range from 19.5 to 23 days. During the worst season, 1860, whalemen went 27.9 days between catches. In the other four seasons, bowheads were captured at intervals of between 13.4 and 17.5 days. Whalemen who set out from Hawaii in March for the Western Arctic between 1858 and 1870 enjoyed respectable and fairly predictable results. Confidence that a sufficient number of whales were available in any given year for the taking likely did not foster the tensions that lead to litigation.[6]

The captains who decided to hunt in the Sea of Okhotsk experienced a run of successful seasons lasting until 1858. While the Western Arctic fleet endured the disastrous season of 1851 and a stretch of poor returns through 1857, whalers in the Sea of Okhotsk generally prospered. In 1852, ships returned from the Sea of Okhotsk with a remarkable average of about 1314 barrels. Even as an increasing number of captains in the mid-1850s opted to avoid the Western Arctic and fish the Sea of Okhotsk, catches in the latter remained impressive. The average barrels of oil per ship in 1854 reached 1000 and remained a respectable 812 in 1856. Yet, to borrow a phrase from another extractive industry, the canary was showing signs of distress. The size of bowheads taken in the Sea of Okhotsk, as measured by the average barrels of oil produced per whale, dropped precipitously from 120 in 1847 and 1848 to 94 for the years 1849 through 1852. The decline, an indication that whalemen were willing to pursue younger and smaller whales as the population fell, continued for the period from 1853 to 1861. While the number of bowheads taken in the mid to late 1850s might have remained steady or – in some seasons – even increased, the average whale produced a mere 68 barrels. Only after the number of whalers in the Sea of Okhotsk dropped after 1861 did the average size of captured bowheads increase slightly to 71 barrels.[7]

In 1858, ships in the Sea of Okhotsk experienced the first in a series of weak hunting seasons that signaled the demise of the fishery over the next decade. Not even the return of a substantial number of Northern fleet vessels to the Western Arctic in 1858 and the following seasons could substantially prolong commercial whaling in the Sea of Okhotsk. The 133 ships in the Sea of Okhotsk

[6] Bockstoce, Botkin, Philp, Collins, and George, "The Geographic Distribution of Bowhead Whales," 1–43.

[7] Whaling statistics for the Sea of Okhotsk have not been as systematically compiled or studied as those from the Western Arctic. The Sea of Okhotsk awaits its John Bockstoce. Prior to his death in 1999, whaling historian David A. Henderson gathered much information on the Sea of Okhotsk which is available at the New Bedford Whaling Museum. For Henderson's calculations on the average oil yields for bowheads in the Sea of Okhotsk, see Box 3, Part 1, #17, David A. Henderson Collection, New Bedford Whaling Museum, New Bedford, Massachusetts (NBWM) #2008.18. The average annual number of barrels per ships cited in the text were drawn and calculated from the following sources: *Friend* (Honolulu), December 17, 1852; *Whalemen's Shipping List and Merchant's Transcript* (New Bedford, MA), January 9, 1855, January 13,1857, and January 12, 1858; and *The Pacific Commercial Advertiser*, Honolulu, December 4, 1856.

in 1858 harvested, on average, only 589 barrels of oil. The average for 125 vessels the following season was – at 587 barrels – nearly identical. The 1860 average of only 437 barrels for 103 ships was the worst year in the history of the fishery and would remain so until 1869 when 14 ships brought home an average of 429 barrels. A steep decline in the number of ships hunting in the Sea of Okhotsk allowed the average to rise in 1861 and 1862. A good season in 1862 – 920 barrels – was followed by another poor year in 1863 with an average just below 500 barrels. The 972 barrel average in 1864 was enjoyed by only a dozen ships and certainly did not mark the return of the fishery to economic health. During the remainder of the decade, no more than 20 ships plied the Sea of Okhotsk in a single season. Results were generally discouraging and in 1870 only one whaler sailed into the Sea of Okhotsk. The 200 barrels taken by the *Monticello* that season did not inspire other captains to renew their pursuit of Okhotsk bowheads. In 1871, the bowheads in the Sea of Okhotsk swam free of commercial predation for the first time since the mid-1840s. They were, thereafter, visited by whalemen only occasionally and always in small numbers.[8]

If, as this chapter argues, a fishery's inconsistent annual yields created tensions at sea which reduced the effectiveness of the traditional means of dispute resolution, 1852 would seem a curious season to generate *Taber v. Jenny*, the first American litigation of a whaling contest since the eighteenth century. At an average catch of over 1300 barrels of oil, 1852 may well have been the most successful season in the history of the Okhotsk fishery. The plaintiff's vessel, the *Hillman*, returned to Hilo on October 24, 1852 with a prodigious 2000 barrels of whale oil. The *Zone*, which had used the cover of fog near the Shantar Islands take a whale killed and anchored by the *Hillman*, reached Oahu that fall with a respectable 950 barrels of oil. Why, then, could this matter not be successfully resolved to the satisfaction of all by the captains at sea or before arbitrators two years later in New Bedford? There are at least two likely reasons why this matter ended up in the Boston courtroom of United States District Court judge Peleg Sprague. The practice of a boat's crew killing a whale and anchoring it for safekeeping until a return to the ship for rendering was feasible was common in the shallow bays of the Sea of Okhotsk where ice often necessitated that the small whaling boats fish where ships could not go. The same conditions that kept the crew from finding or reaching their ship with the whale often sent the men to shore with the hope that the anchored bowhead would not be lost or disturbed. Sperm whalers and bowhead hunters in the

[8] *Friend* (Honolulu), November 8, 1858, December 4, 1858, November 1, 1859, December 1, 1859, November 1, 1860, December 1, 1860, October 19, 1861, November 1, 1861, November 18, 1861, December 2, 1861, January 1, 1863, October 2, 1863, November 2, 1863, and December 1, 1863. *Whalemen's Shipping List and Merchant's Transcript* (New Bedford, MA), January 10, 1865, January 30, 1866, January 22, 1867, January 14, 1868, February 1, 1870, January 31, 1871, and January 19, 1875.

Arctic typically lowered their boats into the water upon sighting a whale. The immediate return to the ship was generally easy to accomplish, eliminating the necessity of anchoring a whale and creating a tempting target for another ship happy to gain oil with minimal exertion. While Ellickson's theory that the switch to bowheads created confusion in customs that ended in litigation is inaccurate, he may well be correct that alterations in hunting practices encouraged by the unique weather conditions and water depth in the Sea of Okhotsk caused a period of confusion that whalemen found – at least initially – difficult to resolve on their own. That *Bartlett v. Budd*, which also involved an anchored whale, was litigated following the successful 1856 season supports the notion that changes in whaling techniques disrupted patterns of dispute resolution. A second reason why the owners of the *Hillman* – despite the success of its 1852 season and voyage – might have decided to expend the energy and money necessary to pursue arbitration and litigation is that the dispute had, as previously explained, become personal.[9]

The other three litigated cases from the Sea of Okhotsk did not concern anchored whales and therefore cannot be explained by a change in whaling practices. Involving the pursuit of live whales, these matters appear to be more likely the product of lean, inconsistent seasons in an unique fishery. With an average oil per ship take of 808 barrels for ships plying the bowhead fisheries of the North Pacific and Western Arctic, 1857 was the last of a run of successful seasons in the Sea of Okhotsk dating back to 1851. Despite the overall positive numbers for the Okhotsk fleet, the fishery was showing the signs of a coming decline in the amount of oil per ship that would mark the final productive years of the grounds. The *Whalemen's Shipping List* of January 12, 1858 noted ominously of the 1857 season that "[o]f the fleet in the Ochotsk Sea, some of the vessels have met with good success, and others have done comparatively nothing." The problems hinted at by the *Whalemen's Shipping List* became apparent the following season. The Okhotsk fleet returned to Hawaii in the Autumn of 1858 with a paltry average of 589 barrels. That season the first of the three litigated matters involving the pursuit

[9] For the oil gathered by the *Hillman* and the *Zone* during the 1852 season, see *Whalemen's Shipping List and Merchant's Transcript* (New Bedford, MA), January 4, 1853. *Taber v. Jenny*, 23 F. Cas. 605; Case File, *Taber v. Jenney*, United States District Court, Massachusetts, Special Court, December 1855, United States National Archives, Northeast Region, Waltham, Massachusetts (National Archives, Northeast Region). For contemporary descriptions of extended periods of whaling from boats in the Sea of Okhotsk, see "Northern Whaling," *Overland Monthly and Out West Magazine* 6, no. 6, (June 1871): 548–54; and Edward Dusseault, "Recollections of Other Days," *Ballou's Monthly Magazine* 49, no. 6 (June 1879): 556–61. Dusseault recounts an 1857 whaling trip to the Sea of Okhotsk. For the results of the 1856 season, see *Whalemen's Shipping List and Merchant's Transcript* (New Bedford, MA), December 4, 1856 and December 11, 1856. The 1856 average for the Sea of Okhotsk was 812.8 barrels of oil per ship. In *Bartlett v. Budd*, the plaintiff's vessel (*Canton Packet*) gathered 1000 barrels and the defendant's ship (*Emerald*) took 500 barrels.

of a live whale by two ships was spawned by an early August dispute between the *Washington* and the *Endeavour*.[10]

Litigated in the United States District Court in Boston as *Bourne v. Ashley*, the conflict began when the *Endeavour* spied a whale at a distance of about a quarter mile. When the *Endeavour* began its chase, Captain Richard E. Wilson was aware that the *Washington* was already in pursuit. The *Endeavour* reached the whale first, but the *Washington* quickly joined the frenzied effort to land a harpoon. The depositions introduced at trial tell a confusing story. It is apparent, however, that both ships hurled numerous harpoons which ultimately drew and employed hand and bomb lances. Judge Peleg Sprague, who ultimately decided the matter in the *Washington's* favor without providing a written opinion, likely found the actions of the *Endeavour* less than honorable. Members of the *Endeavour's* crew testified that one of their officers had attempted to slip unnoticed a harpoon into the dying whale to bolster their claim of affixing the first iron. The scheme unraveled when the furtive strike was detected by the *Washington* and it was later observed that the harpoon had never been attached to a rope. The *Endeavour's* rebuttal that an inexperienced greenhand had failed to attach a rope to the iron which had been thrown early in the chase probably sounded as contrived to Judge Sprague as it reads more than a century later. As with all the other whaling cases, it is difficult to determine precisely why this dispute was litigated. The attempted deception might – as in *Taber* – have turned a conflict that was amenable to successful negotiation into a matter of principle. The economic stakes in this dispute, however, doubtless loomed large. The *Washington* returned to Hawaii in the Fall of 1858 with a mere 600 barrels of oil in its hold. While this was about 11 barrels above the fishery's seasonal average, the whale they lost to the *Endeavour* was – at 175 barrels – exceptionally large and rich in some of the finest bone of that or any other season. The failure to take a single whale that would have turned a poor season into one more in line with the catches of previous years was doubtless made even more galling by the *Endeavour's* dishonorable ploy.[11]

Although the 1863 dispute between the *Oregon* and the *Richmond* occurred at the start of the hunting season before conditions and the availability of bowheads that year was known, Heppingstone and Mammen were both veterans of recent lean harvests in the Sea of Okhotsk. Having experienced the

[10] For the 1857 season, see *Whalemen's Shipping List and Merchant's Transcript* (New Bedford, MA), January 12, 1858. For the 1858 season in the Sea of Okhotsk, see *Friend* (Honolulu), December 4, 1858. *Bourne v. Ashley*, 3 F. Cas. 1002 (1863).

[11] *Bourne v. Ashley*, 3 F. Cas. 1002. Case File, *Bourne v. Ashley*, United States District Court, Massachusetts, Special Court, September, 1866, United States National Archives, Northeast Region, Waltham, Massachusetts (National Archives, Northeast Region). For the statistics on the 1858 season including the hauls of the *Washington* and the *Endeavour*, see *Friend* (Honolulu), December 4, 1858.

difficult seasons of 1858 through 1860, Heppingstone and Mammen were well aware that the success of 1862 provided few oracular clues for the present season in such an erratic fishery. Perhaps it was the grim results of the 1863 whaling season in the Sea of Okhotsk that caused Heppingstone to seethe all summer about the bowhead denied him in May and made Mammen unwilling to reach a compromise even in the face of threatened litigation. While bowheads were reported as abundant in the Sea of Okhotsk that summer, the weather – rough and foggy – frequently prevented whalemen from lowering their boats. Forest fires along the western shore of the Sea of Okhotsk added to the whalemen's woes that season. Speculation that the Russians, ever anxious to drive Americans and Europeans out the Sea of Okhotsk, had started the fires was never substantiated. The net effect of the weather and fires was, as *the Honolulu Commercial Advertiser* reported, "to put a check to whaling in the Bays and near the shores, which are the favorite haunts of the whales." The fourteen vessels – for which reliable figures are readily available – that returned to Honolulu that autumn from the Okhotsk fishery arrived with an average of only 491 barrels of oil. Significantly, Captains Heppingstone and Mammen were among those with little to show for their efforts. Mammen's *Oregon* garnered only 480 barrels. For Heppingstone, the results were even bleaker: a mere 160 barrels of oil.[12]

With the Okhotsk bowhead fishery in its final years, the 1867 season was, at an average of 665 barrels per ship, fairly successful. Yet, as was typical in the Sea of Okhotsk in its period of decline, some ships fared quite well and others did not. With only fourteen ships in the Sea of Okhotsk that season, the *Hercules* and the *Rainbow* should have had plenty of room to fish without interference. Instead, those vessels engaged as previously discussed in a dispute ultimately decided by Judge Lowell as *Swift v. Gifford*. The intensity of that conflict and the disinclination of the *Rainbow* to accede to the request of the *Hercules* that the whale be evenly divided was likely – at least in part – the product of the dismal seasons being suffered by the competing ships. The *Hercules* returned to Hawaii that Autumn with only 290 barrels of oil to show for its effort. That the *Rainbow* managed to take a mere 140 barrels of oil that season – including that rendered from the disputed bowhead – does much to explain the desire of Captain Baker to hang onto what little oil he had.[13]

[12] *Friend* (Honolulu), January 1, 1863, October 2, 1863, November 2, 1863, and December 1, 1863. For the report from the *Honolulu Commercial Advertiser* on the 1863 season in the Sea of Okhotsk, see *Whalemen's Shipping List and Merchant's Transcript* (New Bedford, MA), February 2, 1864. *Heppingstone v. Mammen*, 2 Haw. 707.

[13] Case File, *Swift v. Gifford*, United States District Court, Massachusetts, March, 1872, United States National Archives, Northeast Region, Waltham, Massachusetts (National Archives, Northeast Region). For the total returns from the 1867 season including the *Hercules* and the *Rainbow*, see *Whalemen's Shipping List and Merchant's Transcript* (New Bedford, MA), January 14, 1868 and January 28, 1868. The size of the whale in *Swift* is not stated in the opinion or the case file.

III

Consideration of the weather and ice conditions in the Sea of Okhotsk bowhead fishery and the legal disputes it spawned suggests several conclusions and a few caveats. While it is instructive to draw a connection between success and failure in a whaling season and the pursuit of litigation, the correlation should not be overstated. Many – indeed the vast majority – of bad seasons did not send whalemen to the courtroom. This should remind us that people do not simply pursue a resource or take a dispute to court because they are driven by some sort of innate economic imperative. The filing of a lawsuit is the final product of an ongoing series of actions, reactions, decisions, and financial considerations beginning with the initial dispute or confrontation. The origin of some litigation may ultimately be traced to the fact that some captains were simply more contentious than their colleagues. Disputes were likely subject to escalation at sea given the personalities and prior history between participants and the stresses of particular conditions at the time of the conflict. Perhaps John Heppingstone was correct that Captain Mammen was, whatever negative qualities were intended, "a damned Dutchman."[14]

The decision to litigate was ultimately made by owners at a distance from passions inflamed in the competitive chase for a whale. Given the lack of litigation, it is apparent that whatever the personal factors and economic calculations of the cost of bringing an action versus the prospects for recovery, the cooler heads of the counting house almost always prevailed over bad feelings engendered at sea over an allegedly stolen whale. For every Henry Taber who was motivated to file an action after obtaining evidence of Captain Parker's lies and attempt to fix the arbitration hearing, there were likely many businessmen who decided to forego what they believed was a viable action because they simply did not wish to expend the time and suffer the aggravation. Economic hardship was one of many factors determining when whaling disputes were brought to court. The widely fluctuating conditions in the Okhotsk Sea produced inconsistent results between ships and from season to season. This likely put captains and crews on edge which, in turn, made them more likely to fight with other ships over contested whales. If the season turned out poorly or a rival violated the rules of honorable behavior in a particularly egregious manner, the dispute was more likely to follow the ships back to port and send the owners to court. During successful seasons a captain could perhaps be a bit magnanimous, optimistic that other whales were available for the taking. Why waste time arguing when there were boats to be lowered and whales to kill? In a poor season – with the prospects of finding other whales grim – it was well worth the effort to engage in an extended discussion, hoping to perhaps convince a rival captain to split a whale.

[14] Arthur F. McEvoy, *The Fisherman's Problem: Ecology and the Law in the California Fisheries, 1850–1980* (Cambridge: Cambridge University Press, 1986), 13.

8

An untragic commons

Whalemen were, above all else, extremely good at killing whales. Whales were, on the micro level, formidable foes and whaling was an extraordinarily dangerous occupation. On a grander scale, however, whales – be they sperm whales or bowheads – were killed in the nineteenth century at a rate that far outpaced their ability to reproduce. The stories told about commercial whaling before the turn of the twentieth century often have the air of inevitability. The whales, it seems, never had a chance against a determined predator who did not care what they were doing to the stocks of whales worldwide. Viewed in this manner, nineteenth-century whaling appears to be a classic example of Garrett Hardin's tragedy of the commons. Competitively hunting a commonly owned natural resource, whalemen – like those who pursued beaver, bison, and fish – killed their prey at unsustainable rates, unable to resist the lure of short-term higher profits. Individual greed leads tragically, in Hardin's estimation, to collective destruction of a resource that might, with restraint, have been preserved for the short- and long-term profit of all. There is much about this traditional story that is accurate. Sperm whale and bowhead stocks were substantially reduced and grounds which once abounded with whales were emptied by human predation. Yet, closer attention to nineteenth-century whaling reveals that it was not the sort of commonly owned, competitively utilized tragedy imagined by Hardin and those who have appropriated his ideas to cover a myriad of collective action problems. Using Hardin's definition of tragedy, there was, in fact, little which was tragic about nineteenth-century whaling.[1]

[1] Garrett Hardin, "The Tragedy of the Commons," *Science* 162 (December 13, 1968): 1243–8. For other works dealing with the nineteenth- and twentieth-century slaughter of animals and fish, see, for example, Andrew C. Isenberg, *The Destruction of the Bison: An Environmental History, 1750–1920* (New York: Cambridge University Press, 2000); Arthur F. McEvoy, *The Fisherman's Problem: Ecology and the Law in the California Fisheries, 1850–1980* (Cambridge: Cambridge

Hardin saw the tragedy in how people deal with collective action problems not so much in the unfortunate results that ensue, but in the inevitability that people will chose to destroy a resource in exchange for individual, short-term gain. It is the collective inability of those engaged in pursuit of finite resources on a commons to restrain themselves from maximizing their own profit that dooms the resource. There is nothing tragic, in Hardin's telling, about herders adding animals to fields that can accommodate more hungry ungulates. The tragedy arises when the herders are faced with the fact that their field is being overgrazed. Hardin indicates that it is at this point that herders must consider the utility of adding another animal. The rational herder concludes that the personal benefit that will accrue outweighs the deleterious consequences of that extra mouth that is shared by all. The point is that the herder is perfectly aware of both the personal positive of adding an animal and the collective negative of doing so. Hardin writes: "Adding together the component partial utilities, the rational herdsman concludes that the only sensible course for him to pursue is to add another animal to his herd. And another; and another. ... But this is the conclusion reached by each and every rational herdsman sharing a commons. Therein is the tragedy. Each man is locked into a system that compels him to increase his herd without limit–in a world that is limited." The tragedy is in the rationality of adding to the herd even after the pros and cons – what Hardin calls the "component partial utilities" – are considered.[2]

American whalemen in the first half of the nineteenth century, unlike Hardin's herders, did not recognize that their individual decisions to take as many whales as possible posed a danger to whale stocks or the long-term economic success of the industry. Prior to about 1850, whalemen saw little need to seriously think about preserving whale stocks. Religion and their limited understanding of science assured them that whales, as a species, were bountiful and therefore impervious to the limited operations of less than a thousand ships with thirty man crews in vast oceans. The problem for whalemen was always to locate where whales – in their effort to escape hunters – had found refuge. That individual whalemen might take young whales and diminish the number of animals in a particular location was more nuisance than serious threat to the industry. What, from the perspective of the pre-1850 American whaleman, was tragic about an important branch of the national commerce supplying needed products which were beneficial to society? By the time whalemen recognized in the 1850s and 1860s that whale stocks could be damaged by human predation or that bowheads were escaping into icy waters where they could not be followed, coal oil and then petroleum had rendered whale oil a product with

University Press, 1986); John Coleman, *Vicious: Wolves and Men in America* (New Haven: Yale University Press, 2004), and Robert Paddle, *The Last Tasmanian Tiger: The History and Extinction of the Thylacine* (Cambridge: Cambridge University Press, 2000).

[2] Hardin, "The Tragedy of the Commons," 1244.

a limited market. Where was the tragedy in satisfying the diminishing demand for whale products when all agreed that petroleum had saved the species?

I

Herman Melville's discussion in *Moby-Dick* of the possible extinction of whales assured readers that the great creatures were, as a species, immortal. He rejected the comparison with the fate of the disappearing bison with a bit of simple math. Bison, or buffalo as he called them, had been subjected to a level of hunting that far exceeded anything possible in the pursuit of whales. A crew of thirty whalemen on a voyage of four years would consider themselves extremely fortunate to return home with the oil of forty whales. A like number of "moccasined men" could in the same time period slaughter 40,000 bison. Yet, as Melville conceded, sperm whales and bowheads could no longer be found in some of their earlier haunts. This was not a sign of a coming extinction or even a diminishment of numbers, but an adaptation to human predation. Sperm whales having eschewed their previous habit of swimming in widely dispersed small pods, now find safety in "immense caravans." The increasing length of voyages in pursuit of sperm whales was, Melville explained, a result of the decreasing number of habitats employed at any given time. In response to similar pressures bowheads had retreated into icier waters beyond the range of even the most determined whalemen.[3]

While Melville's argument that whale stocks were not significantly declining suggests, at first glance, a profitable long-term future for commercial whaling, there was in his analysis a significant note of caution to anyone pondering the industry's prospects. Melville explained that the salvation of bowheads could be found in their ability to retreat to "Polar citadels" forever beyond the reach of men in boats. Having noted that whalemen were at present passing through the Bering Strait and "into the remotest secret drawers and lockers of the world," Melville revealed – without explicitly saying so – that the industry was approaching or had reached its geographical limits. Bowheads, it seems, had almost reached their sanctuary. Melville's argument concerning the ability of sperm whales to survive – in the absence of an icy barrier – by virtue of their numbers and the historic example of elephants was far less convincing, but the future of whaling was, as whalemen recognized by the 1850s, with the oil and bone laden bowhead and right whales. Published in 1851, Melville's discussion of whale stocks in *Moby-Dick* neatly straddled and contained elements of both earlier and later views of the future of commercial whaling in the face of obvious changes in cetacean populations. Prior to 1850, most American whalemen saw whales as impervious to extinction. Industrious whalemen would always be able to find a profitable new species or fishery. The fear that

[3] Herman Melville, *Moby-Dick or, The Whale* (New York: Penguin Books, 2003), 502.

bowheads would ultimately retreat beyond the area where whalemen could profitably or logistically follow was in the 1850s coupled with the growing realization that new sources of illumination and lubrication would soon prevail. Melville, close to the chronological tipping point between the two visions of the future of commercial whaling, seemed to have his heart in the past, but his head in the future.[4]

Several examples from the 1830s demonstrate the confidence of American whalemen and the educated public as to the health of the industry and the natural resource upon which it depended. Charles W. Morgan, a successful whaling agent and prominent New Bedford civic leader from the 1820s until his death in 1861, presented a Lyceum lecture in 1830 entitled "The Natural History of the Whale." Morgan, preaching to the converted, explained that the oil and bone of whales were "gifts from the ocean" earned by the brave whalemen who force "the Monarch of the deep to yield up his life & his light." It was also hardly a revelation to his audience when Morgan discussed the global economic importance of the industry. After explaining that whale products were indispensable and not equaled or successfully imitated by any "art or science," Morgan turned to the issue of whale stocks. Concern that the very success of the industry would imperil its future was not new. Observers since at least the middle of the eighteenth century had occasionally remarked that overhunting might decimate the species. The pursuit of immature whales – whether for their oil or as a means of securing concerned parents – was cited as a particularly destructive practice. Morgan's response to the issue of whale population was dismissive. The growing length of whaling voyages – a possible sign of dwindling stocks – was the result, Morgan revealed, of cetaceans seeking safer waters. Facing only the minor inconvenience of slightly longer trips, ships continued to return with full holds. The real danger to New Bedford whaling was the increased competition from other American ports. Morgan feared that prices would collapse with a market awash in whale oil.[5]

[4] Melville, *Moby-Dick*, 501, 503.

[5] For a brief discussion of Charles W. Morgan's life and edited excerpts from his Lyceum address, see Dan Bouk and D. Graham Burnett, "Knowledge of Leviathan: Charles W. Morgan Anatomizes His Whale," *Journal of the Early Republic* 28 (Fall 2008), 433–66, 441, 444, 446. For additional information on Morgan, see Zephaniah W. Pease, George A. Hough, and William L. Sayer, *New Bedford Massachusetts: Its History, Industries, Institutions, and Attractions* (New Bedford: Mercury Publishing Company, 1889), 110, 175, 237, 245–6; Lance E. Davis, Robert E. Gallman, and Karin Gleiter, *The Pursuit of Leviathan: Technology, Institutions, Productivity, and Profits in American Whaling, 1816–1906* (Chicago: University of Chicago Press, 1997), 105, 107, 403–4, 521. For Morgan's involvement in the abolitionist movement, see Kathryn Grover, *The Fugitive's Gibraltor: Escaping Slaves and Abolitionism in New Bedford, Massachusetts* (Amherst: University of Massachusetts Press, 2001), 102, 112, 181, 201, 281. For concerns about whale stocks as early as the middle of the eighteenth century, see William Douglass, *A Summary, Historical and Political, of the First Planting, Progressive Improvements, and Present State of British Settlements in North-America* (Boston: Rogers and Fowle, 1749), 297. Douglass stated: "It is not easily to be accounted for, that Whales do not in Course of Years

While Morgan quickly refuted concerns of declining whales stocks in his 1830 speech, he struck a less confident tone when he delivered a revised version of his address in 1837. In the intervening years, the size of the American fleet and the volume of oil delivered to market increased substantially. With a rising demand throughout the decade, the fall in prices Morgan feared in 1830 had not occurred. Yet, Morgan noted in his 1837 remarks that ships returning home with a full hold were, of late, an "exception to a general rule." If Morgan allowed a note of concern to pass his lips, others harbored no doubt as to the future of whaling and the supply of cetaceans available for the taking. An 1834 appraisal in *The North American Review* of Scoresby's *An Account of the Arctic Regions* and other books on whaling was typical. "That a squadron of 700 vessels scour every sea and bay, in the eager and unremitted pursuit, without exterminating or apparently diminishing the species, leaves us," the reviewer opined, "to wonder at the exhaustless resources of nature."[6]

This belief in a limitless supply of whales was vigorously supported by a retired whalemen writing to the *Nantucket Inquirer* in response to an 1834 article describing the coming extinction of hunted cetaceans. The retired captain, who chose to remain anonymous, assured his readers that sperm whales in the Pacific Ocean would never become extinct or even be diminished to a degree which would be noticed by their hunters. Numbering in the millions, Pacific sperm whales were likened to the wild cattle of the "great plains of Buenos Ayres" which were not diminished by the prodigious annual harvest of hides. If a limited stretch of land could hold an inexhaustible supply of cattle, the vast expanse of the Pacific Ocean could certainly serve as an eternal home to a boundless number of sperm whales. Compared to the magnitude of the Pacific and the quantity of whales, a mere fleet of 700 vessels was clearly incapable of erasing an entire species. Indeed, the captain offered, "I am further of opinion that where there is one whale killed by the hands of man, more than a thousand expire in the ordinary course of nature." The aging mariner was even unwilling to concede that hunting had driven whales from their traditional haunts. What

become scarcer; considering the continued great Slaughter of them by the Whaling Nations, they bring only one Calf at a Time after many Months of Gestation, whereas other Fish spawn Multitudes." See also, Samuel Ward, *The Modern System of Natural History: Containing Accurate Descriptions, and Faithful Histories, of Animals, Vegetables, and Minerals* (London: F. Newberry, 1775–1776), 9:26. Eighteenth-century writers also commented upon the decreasing size of whales as evidence that at least species of greater size were being destroyed. Oliver Goldsmith, *An History of the Earth, and Animated Nature* (London: J. Nourse, 1779), 6:187–8. For the issue of hunting young whales and targeting females, see Francois-Alexandre-Frederic La Rochefoucauld-Liancourt, *Travels through the United States of North America, the Country of the Iroquois, and Upper Canada, in the Years 1795, 1796, and 1797* (London: R. Philips, 1799), 1:491.

[6] Bouk and Burnett, "Knowledge of Leviathan," 457. For statistics on American whaling in the 1830s, see Davis, Gallman, and Gleiter, *The Pursuit of Leviathan*, 6, 370–80. "The Whale Fishery," *The North American Review* 38, no. 82 (January, 1834): 84–115, 115.

whalemen sometimes experienced as a troubling shortage of whales in previously well-stocked waters, was, in fact, the result of seasonal variations in the currents driving food into different sections of the sea. Although he limited his remarks to the sperm whales he had hunted, the captain certainly implied that the same principles applied to other cetaceans.[7]

While it would be easy to dismiss Morgan and others in the 1830s as industry boosters stubbornly refusing to recognize that the process of stock degradation might lead to extinction of bowheads or sperm whales, their continued optimism as to the future of whaling was not entirely without support in the scientific community. As recently as the 1780s, Thomas Jefferson could confidently state in *Notes on the State of Virginia* that the "economy of nature" was such "that no instance can be produced of her having permitted any one race of her animals to become extinct; of her having formed any link in her great work so weak as to be broken." Jefferson's statement that no evidence of extinction had been discovered was curious given his avid interest in the mastodon teeth and bones then being unearthed in Kentucky. Rather than accept – as some scientists did – that they were the remains of an extinct animal, Jefferson believed that they belonged to some sort of elephant species previously undiscovered that, quite likely, would be discovered in the exploration of the vast American interior. Nature was complex, but could be understood by recognizing where all forms of life – from the least significant organism to man – fell on what was called the "great chain of being." It was this vision of the universe that Jefferson sought to defend when he indicated that the extinction of an animal would constitute a broken link in the economy of nature.[8]

The prospect that an animal had once existed, but was now extinct posed a serious challenge to the belief that the universe was as the creator intended. Natural theology in the eighteenth century saw in creation evidence of a benevolent God who had generously provided for all creatures. If animal extinction was possible, perhaps the world was not as plentiful and stable as theologians and scientists supposed. Yet, by the end of the eighteenth century, evidence was clearly mounting in the fossil record of animals that had never been observed. In the 1790s, Georges Cuvier demonstrated to the satisfaction of most scientists that that the fossils of mammoths and mastodons were distinct from any known species of quadruped. Cuvier's belief that these fossils

[7] *Nantucket Inquirer* article reprinted in the *New-Bedford Mercury*, March 14, 1834.

[8] Thomas Jefferson, *Notes on the State of Virginia*, ed. David Waldstreicher (Boston: Bedford/St. Martin's, 2002), 116. For the "great chain of being" and Jefferson's view on science and extinction, see Lester P. Coonen and Charlotte M. Porter, "Thomas Jefferson and American Biology," *BioScience* 26 (December 1976): 745–50; C. Edward Quinn, "Thomas Jefferson and the Fossil Record," *Bios* 47 (December 1976): 159–67; Sidney Hart and David C. Ward, "The Waning of an Enlightenment Ideal: Charles Wilson Peale's Philadelphia Museum, 1790–1820," *Journal of the Early Republic* 8 (Winter 1988): 389–418; and Stephen M. Rowland, "Thomas Jefferson, Extinction, and the Evolving View of Earth History in the Late Eighteenth and Early Nineteenth Centuries," *Geological Society of America Memoirs* 203 (2009): 225–46.

were evidence of extinction was not the only possible explanation. That Cuvier was correct that mastodons were extinct has tended to obscure the degree to which other theories remained scientifically viable in the first decades of the nineteenth century. Jean-Baptiste Lamarck saw in the same fossils evidence that the fossilized species had, over time, mutated into the present day elephant. Lamarck found Cuvier's belief that some sort of cataclysmic event – such as the Biblical flood – had wiped out a species inconsistent with his notion of a benevolent God. Instead, Lamarck posited that all species tended over time to grow larger and become more complex. This natural tendency to mutate which Lamarck recognized as progress did not proceed unhindered as if in some sort of laboratory setting. Environmental factors also forced species to change and adapt to new conditions. It was this environmental catalyst of change that was largely responsible, in Lamarck's vision, for the evolutionary change in a single species from the fossilized mastodon to the living elephant. Eschewing extinction and random change, Lamarck preserved the eighteenth-century cosmology of a world moving toward perfection in the progression from simple to complex species and in the development of the human race. Lamarck did, however, concede the possibility that men could intervene in this natural progression and alter the environment that might lead to the extinction of relatively simple species. Larger, more complex species and those that inhabited remote areas – such as whales – were deemed beyond the ability of humans to wipe out.[9]

When Lamarck went to his grave in 1829, it marked – at least symbolically – the passing in the scientific and intellectual communities of resistance to extinction. Extinction as a scientific fact did not, however, immediately gain wide popular acceptance. As late as 1849, *Blackwood's Edinburgh Magazine* believed it necessary in a review of a book chronicling the demise of the dodo to inform its audience that extinction of a species was possible. The idea that a generous God would allow any part of creation to be destroyed remained contrary to how many Christians viewed the world. If Anglo-Americans were slow to embrace the possibility of extinction for land animals, the notion that humans could somehow diminish the bounty of the ocean by fishing and whaling was even more difficult to grasp. Prior to the first real effort to

[9] For the issue of extinction, see Sandra Herbert, "The Darwinian Revolution Revisited," *Journal of the History of Biology* 38 (Spring 2005): 51–66; Ernest Mayr, "Lamarck Revisited," *Journal of the History of Biology* 5 (Spring 1972): 55–94; Richard R. Yeo, "The Principle of Plentitude and Natural Theology in Nineteenth-Century Britain," *The British Journal for the History of Science* 19 (November 1986): 263–82; Samuel T. Turvey and Anthony S. Cheke, "Dead as a Dodo: The Fortuitous Rise to Fame of an Extinction Icon," *Historical Biology* 20 (June 2008): 149–63; George Gaylord Simpson, "Extinction," *Proceedings of the American Philosophical Society* 129 (December 1985): 407–16; Pascal Tassy, "Lamarck and Systematics," *Systematic Zoology* 30 (June 1981): 198–200; John Humphreys, "Lamarck and the General Theory of Evolution," *Journal of Biological Education* 30 (Winter 1996): 295–303; and Frank N. Egerton, "Studies of Animal Populations from Lamarck to Darwin," *Journal of the History of Biology* 1 (Autumn 1968): 225–59.

systematically study the ocean and its floor at mid-century, the sea seemed limitless. Lamarck observed in 1809 that creatures in the ocean "are protected from the destruction of their species by man. Their multiplication is so rapid and their means of evading pursuits or traps are so great, that there is no likelihood of his being able to destroy the entire species of any these animals." Echoes of Lamarck can be heard in the 1845 observations of Charles Wilkes. Wilkes, an American naval officer charged by the government with conducting a survey of the South Pacific in assistance of whaling and other commercial interests, declared the speculation of declining sperm whale stocks unfounded. Sounding much like Lamarck and the retired whaling captain who corresponded with the *Nantucket Inquirer*, Wilkes admitted that whales had become more difficult to catch, but posited that this was not the product of any diminishment in the number of sperm whales. "On an average, it requires fifty whales to fill a ship, and it would therefore take about five thousand whales annually, to supply the quantity of oil that is imported. This would appear," Wilkes continued, making a leap revealing how little was known in the mid-nineteenth century about whale stocks, "but a small proportionate number, if these animals were as prolific as our herds on shore, when it is considered that they have a feeding-ground of twenty millions of square miles."[10]

The 1850s were a period of transition in attitudes about the sustainability of whaling. The 1850 publication of Henry Cheever's *The Whale and His Captors* marked one of the first statements by an American that human predation would lead to the ultimate extinction of whales. Cheever, a minister who had gone to sea on two whaling voyages in the early 1840s to improve his health, wrote a number of widely read accounts of his experiences at sea and as a missionary in Hawaii. Citing accounts of Wilkes and others of successful attacks on sperm whales by killer whales and other "serpents," Cheever offered that with the combination of natural enemies and "its predatory human enemy, the great mammoth of ocean seems doomed to extinction." Cheever's perspective on the industry was, to be sure, far from typical. As a Christian minister, his focus was often more on the immortal souls of the crew than their professional skills or the bottom line of the vessel's owners. The failure of captains to enforce the Sabbath and the resulting immorality of crews led Cheever to doubt the lawfulness and morality of the entire industry. Cheever admitted that he had come to sympathize with the right whale and "how quietly it grazes through the great

[10] For the discussion of the dodo in *Blackwood's Edinburgh Magazine*, see Turvey and Clark, "Dead as a Dodo," 157. Nineteenth-century scientific exploration of the ocean is detailed in Helen M. Rozwadowski, *Fathoming the Ocean: The Discovery and Exploration of the Deep Sea* (Cambridge: The Belknap Press of Harvard University Press, 2005). For Lamarck's quotation, see W. Jeffrey Bolster, "Opportunities in Marine Environmental History," *Environmental History* 11 (July 2006): 567–97, 574. Charles Wilkes, *Narrative of the United States Exploring Expedition. During the Years 1838, 1839, 1840, 1841, 1842* (London: Wiley and Putnam, 1845), 5:493. Wilkes and his mission are discussed at Granville Allen Mawer, *Ahab's Trade: The Saga of South Seas Whaling* (New York: St. Martin's Press, 1999), 190–203.

pasture-ground which God has ordained for it and fitted so well to be its home." Unlike previous literary visitors to whaling vessels, Cheever's introduced – in addition to thrilling accounts of the hunt – the suggestion that the slaughter of such noble creatures constituted a cruel infliction of needless pain. While Cheever was certainly ahead his time in offering the suffering of whales as an argument against the industry, he was very much in the mainstream in linking the future of whaling to the rise of viable alternative illuminants and lubricants. Cheever – in keeping with his missionary aims – was loathe to see the immoral environment on whaling vessels maintained if there were other products capable of serving the needs of society. Other observers, of course, viewed the issue of other oils from a strictly economic perspective.[11]

II

In the eighteenth century, vegetable and animal oils were the only practical fuels for burning in lamps. Whale oils were considered the most reliable and cleanest burning of the available options. The voracious demand in the first half of the nineteenth century for cheap, safe, pleasant, and effective illuminants spurred the search for products to compete with whale oils. Some of the early challengers such as camphene – a distillation of turpentine and alcohol – captured a

[11] Henry T. Cheever, *The Whale and His Captors; or, The Whaleman's Adventures, and the Whale's Biography, as Gathered on the Homeward Cruise of the "Commodore Preble,"* (New York: Harper & Brothers, Publishers, 1850), 125–6. For details on Cheever's life, see Keith Huntress, "Melville, Henry Cheever, and 'The Lee Shore,'" *The New England Quarterly* 44 (September 1971): 468–75. For an interesting contemporary review of Cheever's work by Edward Forbes, a Professor of Natural History at the University of Edinburgh, which stresses the inevitability of extinction as driven by an insatiable demand for whale products, see Edward Forbes, *Literary Papers by the Late Professor Edward Forbes, F. R. S.* (London: Lovell Reeve, 1855), 145–62. Other popular accounts of whaling voyages include, Thomas Beale, *Natural History of the Sperm Whale: Its Anatomy and Physiology – Food – Spermaceti – Ambergis – Rise and Progress of the Fishery – Chase and Capture – "Cutting In" and "Trying Out" – Description of the Ships, Boats, Men, and Instruments used in the Attack; With an Account of its Favourite Places of Resort. To Which is Added, a Sketch of a South-Sea Whaling Voyage; Embracing a Description of the Extent, as well as the Adventures and Accidents that Occurred During the Voyage in Which the Author was Personally Engaged* (London: John Van Voorst, 1839); and Francis Allyn Olmsted, *Incidents of a Whaling Voyage. To which are Added Observations on the Scenery, Manners and Customs, and Missionary Stations of the Sandwich and Society Islands* (New York: D. Appleton and Co., 1841). For a discussion of how Melville, Cheever, Beale, and Olmsted thought about whales and their treatment, see Philip Armstrong, "Moby-Dick and Compassion," *Society & Animals* 12 (2004): 19–35. For the rise of organizations dedicated to eradicating the suffering of animals which were originally directed primarily at domesticated creatures, see Isenberg, *The Destruction of the Bison*, 143–46. The attitude of whalemen in this period to suggestions that whaling was cruel can be seen in a satirical letter to the editor of the *Nantucket Inquirer* arguing that chloroform which will be provided *gratis* by the Society for the Prevention of Cruelty to Animals should "be applied to every whale just before the harpoon is let into him, in order that his 'sufferings' may not be 'intolerable.'" *Nantucket Inquirer*, April 10, 1848.

growing share of the market by the 1840s. While camphene burned more brightly and was cheaper than whale oil, it was also – as whaling interests were quick to point out – very volatile. A lard oil produced from hogs was also introduced in the 1840s, but its poor light and strong odor limited its use for reasons of availability and price primarily to the American interior. Gas produced from coal was in use in a number of cities, including New Bedford, by the 1850s. The price of producing gas and establishing the infrastructure for delivery through pipes limited its commercial appeal. Whale oil remained a superior product and so long as its price remained competitive, gas, camphene, and lard oil would damage, but not strike a fatal blow to the economic health of New Bedford's leading industry.[12]

The sudden appearance and almost immediate recognition of coal oil as a clean burning, safe, and cheaper alternative to whale oil dramatically altered the market for lamp fuels in the mid-1850s. Coal oil, or kerosene as it was also known, was produced by the distillation of coal into a dark, tarry liquid nearly identical to petroleum that, with further refinement, burned brightly and without odor. Combined with the rising price of sperm and whale oil as voyages grew longer and less productive, coal oil was – as whalemen quickly recognized – more than just another competitor. In late 1856 and early 1857, a number of popular magazines and newspapers ran very similar articles pointing out the advantages coal oil enjoyed over whale oil in price and supply. Publications ranging from *The Eclectic Medical Journal* to the *Pacific Commercial Advertiser* – a Honolulu newspaper read widely by the American whaling community in Hawaii – placed the matter in stark numbers. In 1855, the American whaling fleet consisted of 635 vessels employing about 15,000 men and capitalized at around $10,000,000. The industry production of 72,640 barrels of sperm oil and 184,015 barrels of whale oil in 1855 was the product of, on average, three years' work. At $1.77 per gallon for sperm oil and $0.71 per gallon for whale oil the total revenues for American whaling in 1855 – excluding whalebone – was $10,500,000. A single coal oil operation in Kentucky, the Breckenridge Coal Company, was with the labor of thirty men able to produce about 20,000 barrels of oil per year. Only twelve coal oil plants of a similar size would be needed to equal the annual oil output of the entire whaling fleet. Given the vast supply of coal in America and the mere $60,000 spent to build the Breckinridge facility, readers of means must have marveled at the money to be made in the nascent coal oil business. Not wishing to let the

[12] For camphene, gas, and lard oil, see Brian Black, *Petrolia: The Landscape of America's First Oil Boom* (Baltimore: The Johns Hopkins University Press, 2000), 19–22; and Eric Jay Dolin, *Leviathan: The History of Whaling in America* (New York: W. W. Norton & Company, 2007), 336–41. For a contemporary discussion of the strengths and weaknesses of the competing illuminants, see T. Allen, "Explosibility of Coal Oils," *Annual Report of the Board of Regents of the Smithsonian Institution Showing the Operations, Expenditures, and Condition of the Institution for the Year 1861*, 37th Congress, 2nd Session, Mis. Doc. No. 77.

meaning of the comparison be missed, the author or authors of these articles also included the somewhat fanciful figures of what the number of men and amount of capital employed in whaling could achieve if set to work producing coal oil. The resulting yearly income would be a staggering $275,000,000.[13]

While the articles read like press releases from the Breckinridge Coal Company or other related concerns – as they might well, in part, have been – the message for the whaling industry was clear and could not be readily contested. During the decade of the 1850s, whale oil's share of the value of illuminants and lubricants produced in the United States fell by two-thirds. The raw material of illumination would, thereafter, come from below the soil of places like Pennsylvania and Kentucky. Although the American whaling industry – given an ongoing demand for baleen and whale oil lubricants – would continue to send ships out to sea, its brightest days were past. Coal oil's position as the next great illuminant was, however, short-lived. Experiments in the late 1850s demonstrated that petroleum could be distilled by the same method used to produce coal oil. When a ready supply of petroleum was discovered in the Oil Creek Valley of Western Pennsylvania in 1859, the technical knowledge for production and an infrastructure for distribution were already in place. Given the equal quality of the products and the higher cost of production, coal oil could never compete with petroleum. Having paved the way for petroleum's ascendency, coal oil never enjoyed the profits envisioned in 1856.[14]

[13] For the process of producing coal oil, see Allen, "Explosibility of Coal Oils." For the early coal oil industry, see Samuel Rezneck, "Coal and Oil in the American Economy," *The Journal of Economic History* 7 (1947): 55–72; John Butt, "Technical Change and the Growth of the British Shale-Oil Industry (1680–1870), *The Economic History Review* 17, New Series, no. 3 (1965): 511–21; and Kendall Beaton, "Dr. Gesner's Kerosene: The Start of American Oil Refining," *The Business History Review* 29 (March 1955): 28–53. The importance of lamp technology in making particular fluids practical and appealing to consumers is discussed in Mimi Sherman, "A Look at Nineteenth-Century Lighting: Lighting Devices from the Merchant's House Museum," *APT Bulletin* 31, no. 1 (2000): 37–43. Publications touting the bright future of coal oil include: "Coal Oil for Illumination and Lubrication," *The U. S. Nautical and Naval Journal* 5, no. 3 (December 1856): 173–6; *Pacific Commercial Advertiser* (Honolulu), February 12, 1857; and J. Milton Sanders, "The Ancient and the Modern Magician," *The Eclectic Medical Journal* 1, Fifth Series (January 1857): 21–7.

[14] The connection of the articles touting coal oil to Breckinridge Coal Company or some other coal based product is unknown, but the pieces in *The U. S. Nautical Magazine* and the *Pacific Commercial Advertiser* make reference to the address of the Breckinridge Oil Company office in New York City where samples of their products can be examined. J. Milton Sanders, the author of the piece in *The Eclectic Medical Journal*, was a chemist who had developed a process in the late 1850s using coke to produce a water gas illuminant. For Sanders, see *The American Gas Light Association. Report of Proceedings* (Providence: Press of E. L. Freeman & Son, 1892), 157–9; and "Gas for Nothing," *San Joaquin Republican* (Stockton, CA), April 17, 1858. For the rapid transition from coal oil to petroleum, see Black, *Petrolia*, 20–1. For the drop in whale oil's share in production of illuminants and lubricants in the 1850s, see Davis, Gallman, and Gleiter, *The Pursuit of Leviathan*, 362.

New Bedford's leading whalemen – recognizing the seriousness of the growing competition provided by coal oil, gas, and other competitors in the 1850s – began by the middle of the decade to seek other investment opportunities. In 1857, Abraham H. Howland, a prosperous ship owner and former mayor of New Bedford, joined with a number of relatives and other scions of the city's elite whaling families to establish the New Bedford Coal Oil Company. With the market for coal oil collapsing, the company began in 1861 to refine petroleum. Examination of the credit reports prepared by the R. G. Dun & Co. reveals the degree to which New Bedford whaling merchants were in economic distress by 1860. Entries after 1858 repeatedly stated that with whaling "going astearn" the subject's worth had dropped. While not all whaling merchants had the financial resources to invest, like Abraham Howland, in a petroleum refinery, many sought other opportunities. Benjamin Howard invested in a flour mill and Matthew Howland purchased real estate in New York. Most maintained some money in whaling, but a few – such as Rodney French – abandoned the industry. Dun reported in August 1861 that what French "owns in whaling is probably not wor much." The next entry for French indicated that he had left New Bedford by 1863 to mine in Colorado.[15]

The leading whaling merchants were not, of course, the only people in the late 1850s who recognized the difficult prospects facing the industry. Newspapers and popular magazines were filled with articles lauding the safety and the low price of coal oil. The corollary message – that whaling's time as an important component of the American economy was finished – was impossible to miss. In 1859, *Ballou's Dollar Monthly Magazine* reported, after characterizing coal oil as "cheap and brilliant in its light," that "the whaling business seems to be gradually superseded." *Friend's Review* detailed the advantages of coal oil and the rising demand for illumination before advising its reader to head the whispers of "The Great Spirit" to "turn your eyes from the depths of ocean to the caverns of the land, and you will find vast and illimitable stores of oil, which have been laid up for the children of this age from the foundation of the earth." There were attempts in publications such as *The Friend*, a Honolulu newspaper that served as an important source of information for participants in Pacific whaling, to push back against the tide of favorable press enjoyed by coal oil. Coal oil, *The Friend* reported in March 1860, was subject to explosions, produced a "noxious odor," and "is said to be very injurious to the lungs." The rearguard action of *The Friend* against coal oil in declaring that

[15] For the New Bedford Coal Company, see Pease, Hough, Sayer, *New Bedford Massachusetts*, *177–8*; *The Derrick's Handbook of Petroleum. A Complete Chronological and Statistical Review of Petroleum Developments from 1859 to 1898* (Oil City, Pennsylvania: Derrick Publishing Company, 1898), 974–6; and *Appleton's Annual Cyclopedia and Register of Important Events of the Year 1901* (New York: D. Appleton and Company, 1902), Third Series, 6: 437. Massachusetts, Vol. 17, p. 435, 437, 451, 467, R.G. Dun & Co. Credit Report Volumes, Baker Library, Harvard Business School.

there was a future – albeit somewhat diminished – for whalemen was, when published, already a moot point. Whether whaling could have fought back the challenge of coal oil – while certainly doubtful – must remain a matter of speculation. The production of American oil wells quickly overwhelmed all competitors after Edwin Drake struck oil in 1859. Within two years the yearly American petroleum output matched the annual output of whaling in its best year and within six years more crude oil was pumped from the ground than American whalemen managed to wrest from the sea for the entire period from 1816 to 1905.[16]

By the early 1860s, it was widely perceived that the discovery of a ready supply of petroleum had saved the whales. The degree to which this view permeated the popular culture can be seen in a cartoon that appeared in the April 20, 1861 edition of *Vanity Fair*. Captioned "Grand Ball given by the Whales in honor of the discovery of the Oil Wells in Pennsylvania," the cartoon depicted whales in formal evening wear making merry with drink under banners celebrating their deliverance.

One banner toasted "The oil wells of our native land may they never secede." Other signs proclaimed "oils well that ends well" and "we wail no more for our blubber." The notion of a petroleum based cetacean salvation was also supported by the author of a lengthy letter to the editor of the *New York Evening Post* chronicling a visit to the oil region of western Pennsylvania. The letter, reprinted on September 29, 1860 in *The Living Age*, provides a paean to the abundance and usefulness of petroleum before ending with the prediction that there is a "good time coming for whales." Even the *Whalemen's Shipping List*, the New Bedford based unofficial industry paper, could not resist the image of the appreciative whale. The July 2, 1861 issue featured a reprint from the *New York Journal of Commerce* which suggested that "the whales themselves will undoubtedly be grateful for the discovery of the oil which is fast superseding that hitherto supplied by themselves."[17]

The decade of the 1850s witnessed a remarkable change in how people viewed the future of whaling. In 1850, it was still possible to view whaling as an industry with a viable future. By the early 1860s, it was universally conceded that whale oil would never regain its position in the illuminant market. A wan note of hope was struck for whaling's future by the *Scientific American* in late

16 "Time's Changes," *Ballou's Dollar Monthly Magazine* (September 1859): 297. "Oil of Coal," *Friend's Review; A Religious, Literary and Miscellaneous Journal* (May 10, 1856): 550. "Whale Oil v. Kerosene," *Friend* (Honolulu), March 1, 1860. Comparative statistics of whale oil and petroleum output can be found in Davis, Gallman, and Gleiter, *The Pursuit of Leviathan*, 363.

17 *Vanity Fair* (April 20, 1861): 186. *The Living Age* (September 29, 1860): 810–2. *Whalemen's Shipping List* (New Bedford, MA) July 2, 1861. The *Whalemen's Shipping List* article is also quoted at John R. Bockstoce, *Whales, Ice, & Men: The History of Whaling in the Western Arctic* (Seattle: University of Washington Press, 1986), 167.

FIGURE 10 "Grand Ball Given by the Whales in Honor of Oil Wells in Pennsylvania." Vanity Fair, 1861.
Courtesy of the New Bedford Whaling Museum

1862 when it offered word of rumors that the flow of petroleum from American wells was slowing. Perhaps given a brief reprieve from hunting, whales might return to their old haunts and the industry could be reborn. What little optimism the article suggested was, however, overwhelmed by its primary message. Entitled "The Innovations of the Age," the thrust of the piece was that as one resource becomes obsolete or is used up a new product or technology is discovered to deliver the necessities and comforts of life. Whale oil – the reader learns without having been explicitly told – is, like the burning of wood for heat, a relic of the past doomed by a superior fuel and the high price of a dwindling resource. The old economy of nature that viewed all creatures having an eternal place in creation was replaced by an economy of nature that saw in the passing of species the path to a perfected nature. Human society – governed by the same principles – was, likewise, advanced by the use of resources until they were no longer available or needed. The God that assured people that they would always have enough whales gave way to a nature that would provide, in combination with human invention, a superior replacement. The 1860 *Preliminary Report on the Eighth Census* explained the recent decline in whales and the resulting increase in lard oil production as an example of "that beneficent law of compensation which pervades the economy of nature, and when one provision fails her children, opens to them another in the exhaustive storehouse of her material resources, or leads out their mental energies upon new paths of discovery for the supply of their wants."[18]

Hardin's herders were tragic because they recognized what they were doing to the land and, yet, they continued to add to their herds. Driven by some sort of innate imperative to make as much money as they could in the short run, they could do nothing to avert the financial collapse of the entire community. Even had they reached an agreement to limit the herd, all participants understood that other herders would cheat because it was in their economic self-interest to do so. American whalemen in the nineteenth century were not tragic in that they did not believe prior to 1850 that they were doing any real harm to the stocks of whales they were chasing. There was no reason in their minds why – with the possible exception of immature whales – they should ever forebear in taking a whale. By mid-century, when the possibility that whale populations were in danger was generally accepted, the sudden rise of alternative sources of oil left whaling a dying industry. There was no longer either an economic reason to preserve whale stocks for future use or a concern that a dying industry might wipe bowheads and sperm whales from the seas. Had American whalemen prior to the 1850s recognized the industry's toll on whale stocks and the need to conserve, their failure to do so would – given their obvious ability to self-regulate disputes – have, indeed, constituted a tragedy of the commons.[19]

[18] "The Innovations of the Age," *Scientific American* (December 1862): 378. The *Preliminary Report of the Eighth Census* is quoted in George W. Shuster, "Productivity and the Decline of American Sperm Whaling," *Boston College Environmental Affairs Law Review* 2 (Fall 1972): 345–57, 347. For the ideas about evolution and science that changed conceptions of the economy of nature in the second half of the nineteenth century, see Thomas Gonderman, "Progression and Retrogression: Herbert Spencer's Explanations of Social Inequality," *History of the Human Sciences* 20, no. 3 (2007): 21–40; Hannah Gay, "Explaining the Universe: Herbert Spencer's Attempt to Synthesize Political and Evolutionary Ideas," *Endeavour* 23, no. 2 (1999): 56–9; Egerton, "Studies of Animal Populations from Lamarck to Darwin," 225–59; and the still useful Richard Hofstadter, *Social Darwinism in American Thought* (Boston: Beacon Press, 1962), particularly 31–50.

[19] It is also possible, of course, to argue that nineteenth-century whaling – as an industry – was tragic in a broader sense because it killed so many whales. Unless one is willing to assert that the loss of a single animal life to meet the needs of human beings is tragic, where does one draw the line between reasonable predation and tragic slaughter? If the line is some measure akin to maximum sustainable yield, it would be difficult even given the state of current science to precisely fix such a point for any particular moment in the nineteenth century beyond which the taking of another whale would be tragic. For a discussion of this issue, see Mark Sagoff, *Price, Principle, and the Environment* (New York: Cambridge University Press, 2004), 16. Should humans have totally abstained from the use of whale oil or should they, as Sagoff facetiously asks, have limited their take in an effort to preserve a "strategic reserve of blubber?"

9

Conclusion

This book, as was stated in the Introduction, is about how whalemen resolved disputes at sea and how courts and legal professionals created the property law of whaling that they memorialized in judicial opinions and legal treatises. In order to understand how nineteenth-century American whaling practices came about, how they changed over time, and how the practices of whalers came to be recognized and adapted by legal professionals such as judges, it has been necessary to pay close attention to the particulars of the story. It mattered who did the whaling, where and under what conditions it was carried out, the sort of equipment employed, the business structure of the industry, and so on. All of the hundreds of details about American whaling in this period make this a very particular story unlike that of any other extractive industry. This is what historians do: they tell very detailed stories that are always unique. There will be moments in any good history where a reader might very well conclude that the matter under discussion shares certain features with something from another time and place. This is all well and good, but it is not the primary goal of a historian. Historians deal in the unique, the *sui generis*. Historians and law-and-humanities scholars for the most part are, as Robert Ellickson has recently observed, humanists who "tend to delight at finding evidence of variations in social practices and in explaining the contingency of outcomes."[1]

There is, of course, another way to tell this story about American whaling. As Ellickson has explained, there are also scholars of a more scientific bent who favor the "universalizing" and "the sweeping hypothesis." These practitioners of law-and-economics – in whose camp Ellickson firmly situates his own work – see in the American whalemen of the eighteenth and nineteenth centuries a fine

[1] Robert C. Ellickson, "The Entitlements of Unallied Hunters After a Sequential Capture," *University of Toronto Law Journal* 63 (2013): 137.

example of how a close-knit community goes about resolving disputes over ownership of a valuable natural resource and how lawyers and judges take those norms and turn them into law. Having learned something about these whalemen, Ellickson and his scholarly brethren can add to their composite explanation of how similarly situated groups settle disputes and allocate property. From this, they argue, it is possible to build legal and economic models of human behavior that are as valid for whalemen in the Sea of Okhotsk as they are for ranchers in twenty-first-century Shasta County, California.[2]

While the divide between the humanist law-and-society crowd and the scientists of law-and-economics scholarship is, indeed, wide, it need not be insuperable. The two approaches have much to offer to their opposites. Writing as one trained as a humanist, my objection to much of the law-and-economics literature is not the use of models such as the Coase Theorem, but rather the ahistoricism of the supporting examples. Even those historians inclined to employ economic models in their work frequently find the grasp of the underlying historical examples offered by law-and-economics practitioners decidedly lacking. It is difficult to see much that is scientific about a model built upon a shallow foundation. Having spent so much time on the particulars of nineteenth-century American whaling it is perhaps time to offer some observations about what these whalemen can tell us about how close-knit communities resolve disputes. It is, in other words, time for a little model building.

I

The law of possession, as Carol Rose and others have made clear, is about the communication of information. People who park on public streets and shovel their cars out of deep snow, to use a favorite example of law professors, do not like to return to find the space taken by another vehicle. Although the dispossessed shoveler has no legal claim on the parking space, she is not without a possible remedy for freeloading behavior of – to borrow a memorable phrase from *Pierson v. Post* – a "saucy intruder." She can communicate to the world or at least the world of drivers that her labor in clearing the space entitles her to that spot of pavement upon her return. Placement of lawn furniture, for example, tells would-be parkers that someone has gone to great effort and that she expects her labor to be honored by their forbearance. Lawn furniture is well suited for the purpose. It is large and visible from a distance, clearly marks the space that is being reserved, and is not of such value that it is a likely target of theft. Most importantly, its message is easily understood by both neighbors and visitors to the neighborhood. Lawn furniture in the context of a snowy street is so easily recognizable as a claim of possession,

[2] Ellickson, "The Entitlements of Unallied Hunters," 137.

in fact, that a shared language or even culture is not required for the message to be accurately communicated.[3]

The world of motorists to which the lawn furniture is designed to speak is large and is made up of first-time visitors to the city and long-time neighbors. Given the need to communicate the message to an audience that shares so little in terms of background knowledge of local practices, there is only so much information that the lawn furniture can successfully convey. The lawn furniture simply tells motorists to not park in the cleared space. It cannot communicate to its intended audience that one may feel free to park in the space so long as the space is surrendered and the lawn furniture returned to its original location by 5:00 PM. Neighbors may understand this rather more complicated message but visitors will not. A particular street or neighborhood could, in fact, create a whole system of signals whereby a blue lawn chair says something quite different from a red one. It is, however, highly unlikely that neighbors will go to the effort to create such a system as long as outsiders also park on their streets. Signs could be posted explaining the meanings attached to various colors and types of lawn furniture, but that would require a great deal of effort and not all drivers read English or whatever language is employed. Drivers anxiously looking for a place to leave their vehicle are also not likely to scan telephone poles for missives concerning the iconography of outdoor furniture.

If, however, a neighborhood was so isolated that the only people driving on its streets were residents, an elaborate system of signals could easily be created to pass on a number of rather complicated messages governing parking. This idea, that the smaller and more tight-knit the community of users the more complicated the message that simple signals can convey, has been widely accepted by legal scholars. It makes sense that a small community could without much cost communicate with one another and quickly inform any newcomers of local practices and expectations. It is for these reasons that Henry E. Smith has confidently proclaimed that "information intensiveness will increase with a group's small size and close-knit nature, as well as the amount at stake."[4]

American whalemen in the Sea of Okhotsk were by all measures in the middle decades of the nineteenth century a very small, close-knit community competitively in pursuit of a valuable resource. If Smith is correct, then they should have created a means of communicating claims of possession that was

[3] Carol M. Rose, *Property and Persuasion: Essays on the History, Theory, and Rhetoric of Ownership* (Boulder, CO: Westview, 1994). *Pierson v. Post*, 3 Cai. R. 181 (1805). For some of the literature on property rights in shoveled parking spaces, see Fred McChesney, "Snow Jobs," *Library of Economics and Liberty* (October 15, 2001), accessed June 29, 2015, www.econlib.org/library/Columns/Mcchesneysnow.html; Richard A. Epstein, "The Allocation of the Commons: Parking on Public Roads," *Journal of Legal Studies* 31 (June 2002): S528–33; and Carol M. Rose, "Possession as the Origin of Property," *University of Chicago Law Review* 52 (1985): 81.

[4] Henry E. Smith, "The Language of Property: Form, Context, and Audience," *Stanford Law Review* 55 (April 2003): 1122.

complex and effective. Whalemen hunting in the Sea of Okhotsk shared, after all, the unique experience of pursuing bowheads in a very limited area. This shared experience and the repeated interactions over the course of a short-hunting season should have given rise to signs that were communicatively rich and closely tailored to the prevailing geography, weather, oceanography, and biology of their prey. Yet, as we have seen, American whalemen in the Sea of Okhotsk never created anything approaching clear norms to govern possession of contested whales. They failed, in fact, to establish a means to communicate claims of possession that all participants agreed upon or even understood.

This failure to create a communicatively rich system of signs and signals to clearly and effectively determine which crew should gain ownership of a whale does not mean that they tried and were unable to do so. There is no question that had whalemen wanted to create and be governed by such a system, they would have succeeded. A strict application of fast-fish, loose-fish, and rules that award a crew that comes upon an anchored whale or an injured whale pursued by a first striker as much of the whale as could be rendered before a valid claim was lodged by another ship would, after all, have decided most disputes. The use of flags signaling when a whale had been struck as was sometimes used in the Greenland fishery and the early American practice of awarding an established share to a crew that answered a call for assistance were well within the ability of a close-knit community of whalemen to effectively employ. Rather, whalemen preferred to operate on the basis of vague standards rather than clear rules.

Carol Rose has observed that laws and customs governing property issues veer back and forth between clear, bright-line rules and guidelines that are more amorphous and situational. These competing approaches which Rose has aptly labeled crystals and mud each offer advantages and disadvantages in terms of resolving a problem in a way that is fair and in the best interests of society. Crystalline rules are rigid and have the virtue of certainty. Participants know how a dispute will be resolved and can thus act accordingly. The disadvantage of such an approach is that the outcomes are on occasion unfair to individuals or counterproductive to the shared interests of the community of participants. In the context of whaling, a fishery that strictly adheres to iron holds the whale, for example, puts all ships on notice that a fleeing whale belongs to the pursuing crew whose harpoon is affixed. Competitors must calculate whether it is worth the effort to join the chase in the hopes that the previous striker will fall so far off the pace that all would concur that it no longer had a claim on the whale. Considered in this light, there is little incentive for other crews to take up the chase. The end result is no dispute – generally a good thing – but also the increased likelihood that a whale will escape capture and survive to be chased another day or succumb to its wounds and decompose before it is discovered.[5]

[5] Carol M. Rose, "Crystals and Mud in Property Law," *Stanford Law Review* 40 (February 1988): 577–610.

Lacking the certainty and clarity of crystalline rules, muddy standards are much more difficult to express and are certainly less predictable. They are, however, malleable and can be made to achieve a level of fairness under particular circumstances impossible under a regime of crystalline rules. Consider the differences between a fishery applying a crystalline form of fast-fish, loose-fish and one where fairness – that muddiest of all standards – is followed. A crew that attaches to a whale and endures a dangerous chase through fields of ice over the course of hours only to have its line draw moments before a competitor lands a harpoon in a dying animal is out of luck under fast-fish, loose-fish. All of their effort and sacrifice is for nothing and a newcomer to the chase who endured little of the struggle takes all of the prize. As Scoresby, Melville, and the American captains who testified in *Heppingstone* make clear there is something deeply unfair about such an outcome. A fairness standard such as Scoresby's "laws of honor" looks at the big picture and asks which crew is most deserving of the whale. While fairness is, in general, a laudable goal and perhaps most in line with Lockean notions of the origins of private property, there is a significant potential negative about a muddy whaling standard. Disputes are more likely – at least in theory – where the answer to most any question of ownership is so dependent upon the particulars of how events played out at sea. Flexibility, in other words, would seem to breed conflict.[6]

Yet, as we have seen, whalemen in the Sea of Okhotsk preferred muddy standards and that in using such principles they were remarkably successful at avoiding violent disputes and litigation. This preference of whalemen runs counter to conventional wisdom as to how close-knit communities are inclined to operate in pursuing a limited resource. Economist Harold Demsetz argued that as a resource in a commons grows scarcer and more valuable rules governing entitlements to that property become more clearly defined. This clarity reduces disputes and sets in motion an orderly market in the buying and selling of the desired resource. The end result is greater efficiency. Rose has, however, shown that rather than a progression toward crystalline rules in the face of scarcity, there is a tendency of allocation schemes to alternate between crystal and mud. Clear rules become too rigid and in the name of fairness some mud is mixed in to achieve the desired result. The uncertainty of the muddy standard gives rise, in turn, to a call to establish clearer, more certain rules.[7]

The examples that Rose provides for this seesawing between crystal and mud suggest that it is communities of users who prefer crystalline rules and courts

[6] Rose, "Crystals and Mud in Property Law," 577–610. C. Ian Jackson, ed., *The Arctic Whaling Journals of William Scoresby the Younger: Volume I, The Voyages of 1811, 1812 and 1813* (London: The Hakluyt Society, 2003), 119–20. Herman Melville, *Moby-Dick or, The Whale* (New York: Penguin Books, 2003), 432–5. For the views of the captains who testified in *Heppingstone*, see Chapter 6.

[7] Harold Demsetz, "Toward a Theory of Property Rights," *The American Economic Review* 57, no. 2 (May 1967): 350–3. Rose, "Crystals and Mud in Property Law," 580–3.

that repeatedly insist on muddying things up when rigidity breeds inequity. In the case of real estate transactions, for example, Rose explains that caveat emptor long prevailed under the notion that it was up to the buyer to investigate the property being purchased and make a reasoned decision to enter into an agreement of sale. Both parties were, in theory, equals who were capable of assessing for themselves the wisdom of the deal. The problem, Rose reveals, arises when purchasers are not professional dealers in real estate. Such buyers often do not know and are not in a position to know exactly what they are getting in return for their money. When a purchaser later discovers that a home was poorly constructed he is left without a remedy so long as the doctrine of caveat emptor prevails. Courts, seeing that the purchaser has been taken advantage of by an unscrupulous builder, create the muddy solution of an implied warranty of habitability. The unfortunate buyer has a remedy but real estate professionals now have to deal with the uncertainty of just what it means for a home to be habitable. In order to build certainty back into the system, waivers of warranties and disclosure requirements are inserted into contracts. Once challenged – as they inevitably are – such attempts to reimpose crystals are put to court review and likely subject to replacement by mud. At the core of Rose's examples is the idea that people who are part of a community who must deal repeatedly with some issue concerning the allocation of property prefer clear standards because it is more efficient and maximizes group welfare.[8]

American whalers obviously did not behave in the manner suggested by Rose and other legal scholars. Given that the sort of disputes whalemen engaged in at sea only involved professionals and none of the amateur interlopers such as homebuyers, close-knit communities of whalemen were very much in a position to dictate the means of dispute resolution. They did not face the threat of an outsider stumbling into their world and placing disagreements routinely in front of equity minded judges. Yet, whalemen believed that their interests were better served by the employment of muddy standards.

II

That American whalemen employed neither clear, bright-line rules nor richly communicative signals to resolve disputes does not indicate that the ways by which they decided whale possession were the product of an unpredictable and chaotic social order. Although whaling practices were, as evidenced by the jumble of explanations offered by whaling captains in the Sea of Okhotsk, remarkably vague, it is in that vagueness that can be found the strength and cohesiveness of the order they created.

Legal scholars have long pondered how property rights came into being. Assuming that at some point in the distant past all things were held in common,

[8] Rose, "Crystals and Mud in Property Law," 577–610.

how – the question naturally arises – did humans come to respect and protect individual rights in private property? Two competing creation narratives, each with their own impressive intellectual pedigrees, have emerged. John Locke believed that at some point people sought to protect things that they had taken or forged from the commons. In order to gain a measure of security and avoid the costly and dangerous necessity of defending that property people entered into agreements and arrangements for mutual defense. It is these arrangements that ultimately multiply and coalesce into government. The rules that ultimately govern the ownership of property are said, according to this particular vision of the origins of property, to be the product of an intentional design. People consciously craft the rules and laws that determine when and how one comes to own that which had previously held in common. For Locke and Thomas Hobbes, it is the coming together of people out of fear for the security of their property and their own safety that allows society to prosper collectively and overcome the tyranny of every person acting selfishly in their own self-interest.[9]

The competing notion as to the origins of private property argues that the system evolved in a spontaneous fashion. David Hume believed that communities develop shared interests and ideas as to how things should be done. It is over a long period of time that rules based on these guiding beliefs and principles come into being as people learn the price of transgression and the benefits of compliance. While Hume shared with Locke and Hobbes the indisputably obvious belief that people, when left to their own devices, tend to do what is best for themselves, Hume differed as to the lesson to be taken from this conclusion. A collection of individuals each opting to act in their own self-interest could create, Hume believed, a social order that was peaceful, harmonious, and that worked to the benefit of all. Unlike the chaos Locke and Hobbes foresaw in a society without a vigorous government enforcement of laws to protect property, Hume was rather more optimistic about the prospects of letting society develop as it will.[10]

The importance of this divide in theories as to the origins of private property far exceeds whatever light it might shine on how our ancient ancestors set about to order and govern society. Rather, it is the guidance that these origin stories provide for contemporary economists, political theorists, and legal scholars in suggesting solutions and approaches to current issues that has assured a seemingly endless stream of articles and books on what might appear to be a rather esoteric debate that can never be definitively resolved. Those who follow in the tradition of Locke and Hobbes tend to see whatever order there is in a society as

[9] John Locke, *Two Treatises of Government*, Peter Laslett, ed. (Cambridge: Cambridge University Press, 1988), 267–302 (Book 2, sections 1–51). Thomas Hobbes, *Leviathan*, Michael Oakeshott, ed. (New York: Touchstone, 1997), 98–112 (chapters 13–4).

[10] David Hume, *A Treatise of Human Nature*. James E. Krier, "Evolutionary Theory and the Origins of Property Rights," *Cornell Law Review* 95 (2009): 150–2. Robert Sugden, "Spontaneous Order," *Journal of Economic Perspectives* 3, no. 4 (Autumn 1989): 95.

a product of communal solutions that are intentionally constructed and enforced. Garett Hardin's formulation of the tragedy of the commons falls within this way of looking at the world. Hardin postulated, after all, that individuals sharing a commons who make individual decisions that maximize their own welfare will collectively drive that commons to ruin. The only way out of this inevitable ruin in the context of a collectively utilized commons is the enforcement of rules designed to check individuals from doing that which they would otherwise do. The solution can be imposed by the government or – according to some theorists – be the product of a less formal means of enforcement such as an ethnic community or a trade organization. The crucial element of any solution according to Hardin and others who follow in the tradition of Locke and Hobbes is that it is consciously crafted to effectively check individuals from acting on their own.[11]

For Hume and his fellow Scots Adam Smith and Bernard Mandeville, however, it was not the conscious design of people that made society work but, rather, the unintended consequences of countless individuals acting in their own best interests. As it was impossible to really know enough to understand what is best for all of society it stood to reason that people could not think or consciously design their way out of social problems. Instead, society worked because of the norms and practices that arise spontaneously and come to gain general acceptance as beneficial for the larger community. It was this so-called natural order that if allowed to regulate itself autonomously would produce a well ordered society. While the notion of a natural order applied in theory to a wide array of human activity, it was in the economic sphere through the free market ideas of Smith that this insight came to cast such an enormous shadow over prevailing theories of proper governance.[12]

Smith's metaphor of the invisible hand has, of course, come to be a sort of shorthand for the belief that free markets work to the benefit of society in spreading wealth to all sectors of the economy. While exactly what Smith had in mind when he used the phrase is the subject of much academic debate, the "invisible hand" generally refers to the power of free market forces to create a desirable social order that far exceeds anything that the poor human mind could create. It is this question of how social order is best created that is at the heart of the divide between proponents of conscious design and what Friedrich Hayek famously termed spontaneous order. Hayek, who readily acknowledged his intellectual debt to Smith, believed that people could not design an effective social order because they simply do not have the capacity to communicate and process all of the minute bits of information required. Fortunately and

[11] Garrett Hardin, "The Tragedy of the Commons," *Science* 162 (December 13, 1968): 1243–8.

[12] Bernard E. Harcourt, *The Illusion of Free Markets: Punishment and the Myth of Natural Order* (Cambridge: Harvard University Press), 128–31 and Steven Horwitz, "From Smith to Menger to Hayek: Liberalism in the Spontaneous-Order Tradition," *Independent Review* 6, issue 1 (Summer 2001), 81–97.

fortuitously, prices as set by a free market serve, Hayek argued, as both an aggregator of all this information and the mechanism by which an effective social order is maintained. That effective social order was built on the maximum effective use of resources that the free market would facilitate. Hayek explained that the orderliness of what a free market could create if allowed to was "the product of the action of many men but . . . not the result of human design."[13]

It was Hayek's faith in the efficiency of free market transactions that provides the intellectual foundation for Ronald Coase's seminal 1960 article "The Problem of Social Cost" and the subsequent formulation of the Coase Theorem. Coase, a colleague of Hayek at the University of Chicago, stated – as set forth earlier in greater detail – that in dealing with externalities such as pollution maximum wealth and the efficient allocation of resources is achieved when parties are left alone to negotiate. The giant caveat in the Coase Theorem is that the market reaches this maximum level of efficiency only in the absence of transaction costs. As Coase recognized and as everyone who has ever commented on the Coase's work makes clear, there are always transactions costs in the real world. There are always costs associated with negotiating, threatening, cajoling, and reaching some sort of resolution. The Coase Theorem is widely accepted in terms of Coase's central insight that regardless of how rights are delegated wealth in the absence of transaction costs will be maximized. The elegant example of the rancher and farmer posed by Coase and largely verified by Ellickson, after all, proved as much. In some respects, the Coase Theorem offers something to both proponents of conscious design and spontaneous order. For believers in spontaneous order it is a sort of metanarrative that explains everything. Even for those who favor conscious design, the failure of scholars to offer much research on what happens in the presence of transaction costs makes Coase's Theorem more of a nuisance than a serious impediment to the business of creating solutions to problems of resource allocation and the like.[14]

There is, as Bernard E. Harcourt has observed, a bias in the work of Coase and his followers that is not really supported by the research. Coase believed as an article of faith that market exchanges were always preferable to government intervention in creating an orderly society. The lesson Coase took from his work was that where transaction costs are low government should back off and let the market work but that where such costs were high government should

[13] Harcourt, *The Illusion of Free Markets*, 128. Horwitz, "From Smith to Menger to Hayek," 81–97. Eugene Heath, "Spontaneous Order and Liberalism," *New York University Journal of Law and Liberty* 1, no. 0 (2005), 65–77. Friedrich A. Hayek, *Law, Legislation and Liberty* (Chicago: University of Chicago Press, 1973), 37.

[14] Ronald Coase, "The Problem of Social Cost," *Journal of Law & Economics* 3 (October 1960), 1–44. Robert C. Ellickson, *Order without Law: How Neighbors Settle Disputes* (Cambridge: Harvard University Press, 1991), 1–3. Harcourt, *The Illusion of Free Markets*, 121–8.

still opt not to intervene because in its inability to properly measure and understand the effect of its actions the state would only make things worse. Much of the law-and-economics literature has, as a result, been directed at facilitating low transaction costs so as to allow the market to get things right. As a consequence, Harcourt argues that there is in Coase a faith in a natural order that is very much in line with Francois Quesnay and Hume.[15]

While Harcourt is correct that there is little empirical evidence to support this faith in a natural order, the experience of American whalemen can perhaps offer some insight into how people create order in the absence of law. To do this we must turn our attention away from the issue of whether whaling practices were welfare maximizing which has generally dominated the discussion. Robert Ellickson, for example, has looked at whaling norms and has seen in them the means by which a close-knit community maximizes their welfare. While welfare is not strictly a matter of profits, the number of whales captured and rendered was, Ellickson makes clear, the primary concern of whalemen in defining what constituted their welfare. Other concerns such as peace at sea were, in Ellickson's telling, prized primarily as conditions necessary for efficient whaling.[16]

There is no question that Ellickson is correct in seeing whalemen as resolute in their desire to take as many whales as possible. Yet, the real value in the successful creation of a means to avoid disputes was not in the increase of the number of whales killed and rendered. Whalers were extraordinarily adept at killing whales and with or without such an efficient way to decide possession contests they would have still killed a tremendous number of whales. In addition, whether fast-fish, loose-fish or iron holds the whale was employed probably made little difference to the industry's bottom line. If we turn, instead, to the orderliness that whaling practices promoted, there is much to learn about why whaling practices grew vaguer over time and how close-knit communities function.

III

Looking at how whalemen resolved disputes over the course of approximately one hundred years is instructive. The English Greenland fishery employed a reasonably clear application of fast-fish, loose-fish. Principles of fairness in the Greenland Sea lurked more on the periphery than at the core of the practice. American whalemen in the eighteenth century and the very early nineteenth century followed and clearly preferred iron holds the whale and made generally effective use of signals to communicate the need for assistance of other boats in return for a set share of the whale. The question that arises is why did whaling

[15] Harcourt, *The Illusion of Free Markets*, 121–50. Coase, "The Problem of Social Cost," 1–44.
[16] Ellickson, *Order without Law*, 191–206.

practices lose some of its sharpness and preference for clear standards and become decidedly more muddy as the industry moved into the North Pacific by the middle of the nineteenth century? One fundamental difference in the nature of the industry over time is that voyages grew in length as the distances from home ports increased dramatically. English whaling in the Greenland Sea was a seasonal affair. Crews departed from English or Scottish ports and returned a few months later. Courts, lawyers, and the ready prospect of criminal charges should a dispute escalate into violence were never far away in terms of distance or time. Similarly, American whaling in the eighteenth century and into the early nineteenth century stayed along the Atlantic coast. It was not until crews headed deep into the South Atlantic and then into the Pacific Ocean that voyages became multi-year affairs.

To see why the distance traveled and the amount of time away mattered and why muddier practices became increasingly appealing to American whalemen the ideas of economist Robert Sugden about the creation of spontaneous orders are essential. Sugden, as his use of the phrase spontaneous order indicates, believes that rules regulating property disputes and other human activities are not the product of a conscious design. They instead evolve spontaneously from repeated interactions and decisions made over an extended period of time. As such, the rules that emerge spontaneously are different from norms in that there is nothing normative about them. They may resemble a statement of proper moral behavior that a society has debated and consciously crafted to settle an issue, but, in truth, they have simply evolved to settle some potential area of conflict. A rule that emerges spontaneously may or may not be particularly wise or rational in an objective sense but it will likely take on that appearance with acceptance over time.[17]

Sugden arrives at this conclusion by examining game theory and challenging the assumptions about human behavior behind the game of Hawk and Dove. Often used to understand how competitors resolve property disputes, Hawk and Dove is a two person game that presupposes that when players make a claim on something – a whale, for example – they have two choices as to how to proceed. They can be a hawk and aggressively assert their claim even at the risk of conflict or they can be a dove and seek a compromise or, in the presence of a hawk, back down and abandon the claim. If one player chooses to play the hawk and the other the dove, the hawk takes the entire prize and the dove gets nothing but, at least, is not harmed. Should both players opt to behave as doves then the prize is likely shared. When two players choose to be hawks the parties both suffer a net loss. Game theory is employed to better understand how people make decisions and it is generally built on the assumption that people behave rationally but without any experience or knowledge about their opponent. Players are simply imagined to reach rational solutions based solely upon

[17] Robert Sugden, "Spontaneous Order," 85–97.

their complete understanding of the rules of the game. Sugden argues that when a game is analyzed in this manner there is no way to reach a rational solution because players have nothing upon which to base their strategy of playing hawk or dove in a particular situation.[18]

What people actually do in deciding upon a strategy, Sugden explains, is act on the basis of shared societal knowledge and assumptions. Sugden uses the example of two cars approaching from opposite directions on a narrow road. Drivers do not in this situation contemplate in strictly rational terms with no experiential or cultural frame of reference what they should do. Instead, they act based on their experiences and each determines that the best course of action is – at least in America – to move to the right. A collision is averted not because moving to the right is an inherently rational or morally correct thing to do in this situation, but because it is what drivers in America know that they should do. The terms prominence and salience are sometimes used to explain how conventions such as move to the right work and allow drivers to coordinate their actions without actually being able to communicate with one another. The most famous example of how coordination without communication works dates from work done by Thomas Schelling in the 1950s. Schelling asked people where they would go in New York City to meet someone when the parties had not agreed in advance to a rendezvous location. That a remarkably large number of people selected Grand Central Station reflects that certain solutions struck people as reasonable based on their experience of being familiar with how their contemporaries in 1950s New York City thought. Shared societal conventions such as this develop in an evolutionary fashion. Were this experiment to be repeated today, the rational solution would likely be different given changes in the station's cultural meaning and its function given the termination of intercity train service to Grand Central Station a generation or so ago.[19]

This concept of prominence is key to understanding Sugden's vision of a spontaneous order. In the game of Hawk and Dove, conventions such as move to the right are essential. Without a convention, the game quickly calcifies with average players choosing to act the hawk a certain amount of the time and adopting a dove strategy during the remaining percentage of encounters. The problem with this result is that it leads to fairly frequent contests between hawks. When two hawks collide, damage in the form of violence, a threat of violence, or some other costly event such as litigation ensues. While this might be the nightmare scenario of a state of nature that Hobbes feared, it is not what happens. Instead, players select their strategy of hawk or dove in a given situation based on prevailing conventions. In a property dispute the prevailing

[18] Sugden, "Spontaneous Order," 85–97. Carol M. Rose, "Psychologies of Property and Why Property is Not a Hawk-Dove Game," in *Philosophical Foundations of Property Law*, ed., James Penner and Henry Smith (Oxford: Oxford University Press, 2013), 272–88.

[19] Sugden, "Spontaneous Order," 89–90.

convention is generally that the party in possession or closest to possession should prevail. Conventions such as prior possession or first to pursue create what Sugden calls asymmetries. Asymmetries are crucial because they tell players which party has the stronger claim to the property. Once parties can assess the strength of their claim relative to a competitor's position they can make a determination whether they should act as a hawk or as a dove. The end result is that there are fewer contests between two hawks and thus less danger of violence, threats of violence, or litigation.[20]

Sugden's spontaneous order fits well with what we know about how whalers operated at sea. Whaling rules, practices, norms, and ultimately laws all share a common feature. In one way or another, they all honor effort in taking a whale or the degree to which a crew has come close to killing a whale. That a preference for possession should be the convention honored is logical. As Sugden and others have observed, Anglo-Americans have long fancied Locke's ideas about how private property comes into existence. When whalers sought to resolve disputes at sea and were faced with a game of Hawk and Dove they based their strategies on the philosophical, moral, and religious views they held and on their knowledge of how their competitors looked at the world. As whalemen in a particular fishery almost always shared these beliefs coordination was relatively easy. Using the rules that evolved, whalemen shared a rough idea as to the strength of their claim to a particular whale. A strong claim would encourage a crew to adopt the strategy of a hawk. The stronger their claim and the weaker that of their competitor, the more likely the whaler could be confident that their opponent would play the dove and concede the whale. Closer calls in terms of competing claims might result in a conflict between two hawks, but generally the crew with the weaker claim or less confidence in their claim would – having tested the resolve or confidence of the other boat and found it greater than their own – back down and concede the whale.[21]

Spontaneous order is also effective in explaining why whaling property dispute resolution practices failed in the Sea of Okhotsk by the middle of the nineteenth century. James Krier has explained that Sugden's spontaneous order works best when the property being sought is fairly abundant. There is, Krier asserts, a cost-benefit relationship upon which Sugden's application of spontaneous order to property disputes is built. The power of Hawk and Dove in the context of an asymmetrical convention to prevent a contest requires that the value of the resource being pursued is less than the cost to a party of losing a hawk-hawk battle. Once, however, the value of the resource is deemed to be greater than the cost of losing a battle with another hawk the strategy of backing down and reverting to a dove in the face of a stronger claim makes less sense. This is, it would appear, what happened in the Sea of Okhotsk.

20 Sugden, "Spontaneous Order," 89–90.

21 Sugden, "Spontaneous Order," 81. Krier, "Evolutionary Theory and the Origins of Property Rights," 154–7.

Whales in the Sea of Okhotsk became scarcer and the prospects of finding other whales nearby grew dimmer making the costs of a contest or even litigation appear more reasonable. It was not just a matter of bowhead stocks in general falling, but the very particular conditions in the Sea of Okhotsk that made whaling there even more of a precarious venture economically.[22]

Finally, Sugden's theory of spontaneous order helps to solve the riddle of why whaling rules became muddier as voyages moved into the Pacific Ocean. Krier has explained that a spontaneous order of the variety conceived by Sugden only illuminates the operation of extremely simple property rules. Complicated rules governing property are the product of intentional design. A spontaneous order, on the other hand, is built on an asymmetry in a convention that is relatively crude. Spontaneous orders simply do not create complex rules. That whaling rules became muddier is likely a reflection of the need to reinforce the close-knit nature of the whaling community as they traveled farther and farther away from legal authority. Whalemen reinforced the power of a spontaneous order to remain orderly by drifting toward a muddier standard.[23]

[22] Krier, "Evolutionary Theory and the Origins of Property Rights," 155–6.
[23] Krier, "Evolutionary Theory and the Origins of Property Rights," 156–7.

Index

Adams, John, 68–9, 74
Addison v. Row, 42–3, 48, 51
Allen, Horace, 124
American Revolution, 74
Amundson Gulf, 137
Andrew Marvel, 49
Angell, Joseph, 59, 61–2
Anthony, Job, 75, 78
Arctic Ocean, 120, 134
Australia, 80
Ayan, 85
Azores, 71

Bailey, Judge, 53
Baker, Nehemiah, 116, 142
Ballou's Dollar Monthly Magazine, 155
Barents, William, 37
Barnstable, 108, 110–11
Barnstable, MA, 65
Bartlett v. Budd, 106, 110, 114, 117, 140
Basque whaling, 37–8
Bayley, Judge, 56
Beaufort Sea, 136–7
Bennett, Captain, 88, 90, 95, 102
Bering Sea, 83, 135–6
Bering Strait, 83, 135, 137
Best, Judge, 51, 53
Blackmer, Seth, 91, 95
Blackstone, William, 44, 51, 63
Blackwood's Edinburgh Magazine, 150
Bockstoce, John, 136–7
bomb lances, 27
Bonney, Charles, 117–18, 120
Bordon, Gilbert, 110
Bourne v. Ashley, 119, 141
bowhead whales, 14, 83, 108, 134–5, 149
  behavior, 18, 79, 137
  diet, 21, 85, 137
  population, 18, 133, 138
  range, 83, 137
  stocks, 144
  yield, 23, 139, 141
bowheads, 2, 18, 21, 23, 25, 28–9, 31, 34, 37–8, 59, 70, 81, 83, 85, 134, 136–8, 140, 146, 158
Bracton, 59–60, 63
Breckinridge Coal Company, 154
Brig George, The (First Circuit case), 112
Brigham, Lincoln, 98
British High Court of Admiralty, 100
Buller, Justice, 36, 42
Buster v. Newkirk, 116, 126

Caerwent, 46, 78
Caledonia, 42
Cambridge, 75
Camelford, Lord, 75
Cameron, J. T., 49
Campbell, John, 51
Campbell, Judge, 48, 51
Campbell, Lord, 42
Campbell, Lord President, 41
camphene, 152
Canton Packet, 109, 113
Cape Ann, MA, 64
Cape Elizabeth, 86

Cape Horn, 72, 74
Cape Navarin, 135
Cape of Good Hope, 81
Carrington, Frederick Augustus, 51, 54
Chambre, Judge, 45, 47, 118
Chambre, Justice, 68
Cheever, Henry, 145
Chile, 75
Chukchi Sea, 137
Churchill, Captain, 91
Cicero, 91
close-knit communities, 4, 8, 13, 75, 132, 134, 162, 164, 168
coal oil, 145, 156
Coase Theorem, 8, 160, 168
Coase, Ronald, 8, 168
common law, 43, 47–9, 57, 59–61, 66, 117–18, 121, 127
communis error facit jus, 48
Cook, Christopher, 89–90, 93, 102, 105
Cook, Edward, 108
Cornbury, Viscount, Edward Hyde, 65
Costa, Antoine, 122, 124
Crosby, Owen, 96
customs, 132
Cuvier, Georges, 149
Cyrus, 77

darting gun, 27
Dartmouth, MA, 72
Dartmouth, Nova Scotia, 80
Davis Strait, 41, 71, 79, 100
Davis v. Sturges, 67, 69
Davis, Lance E., 22–3
Davis, Seth, 67
Demsetz, Harold, 163
Dexter, John, 125
Doane v. Gage, 69, 77, 79
Doane, Joseph, 67
Dodge, John, 117–18, 120
dodos, 150
Dolin, Eric Jay, 62
Drake, Edwin, 156
drift whaling, 65
drogues, 25, 45, 77–8
Dunkirk, 76, 80
Dutch whaling, 38
  legislation, 39, 114

East India Company, 74–5
Eastham, NY, 63
Eclectic Medical Journal, The, 153
Eliot and Pitman, 96–7, 105
Ellenborough, Lord, 51
Ellickson, Robert C., 10, 13, 69, 79, 121, 129, 132–3, 135, 140, 160, 167–8
Emerald, 109, 113
Endeavour, 141
Erie, 91
Erskine, John, 43–4, 48
extinction, 151
  whales, 149, 152

Falkland Islands, 73
fast-fish, loose-fish, 3, 29, 33, 36, 41–2, 45, 49–51, 56, 58, 80–2, 98, 114, 124, 131, 134, 162–3, 168
Fennings v. Lord Grenville, 46, 50–2, 54, 61, 68, 80, 118
ferae naturae, 2, 40, 45, 47, 49, 57, 60–1, 63, 97, 112–14, 116, 118, 120, 126–7
Fisher, Gamaliel, 109
Fleta, 59–60
Folkestone, 100
formalism, 103
Fowler, John, 112
Fraser, James, 88
French, Rodney, 155
Friend, The, 155
Friend's Review, 155

Gage, Lot, 67
Gaius, 40
Galapagos Islands, 45, 75, 80
Gale v. Wilkinson, 42
Gallman, Robert E., 22–3
Geertz, Clifford, 7
German whaling, 38
Gifford, Charles, 116
Gifford, Ephraim, 88, 118
Gillet v. Mason, 127
Gizhignskaya Bay, 84
Gleiter, Karin, 22–3
Gordon, Robert W., 59
great chain of being, 149
Griffin v. Thomas, 65
Griffin, Samuel, 64
Gulf of Alaska, 83
Gulf of Anadyr, 135
Gulf of Shelekhova, 83
Gulf of St. Lawrence, 67, 80
Gulf Stream, 37

Hale, Matthew, 62
Hallock, Frederick, 112
Harcourt, Bernard E., 167
Hardin, Garrett, 13, 146, 158, 166
Harmony, 53
harpoon, 25
harpoons, 33, 64, 77, 79, 91, 99, 113, 116, 119, 128
Harriot, 100
Harriot, The (Admiralty case), 100
Hawaii, 92, 134, 151
  legal system, 121
Hawk and Dove, 172
Hayek, Friedrich, 167
Heppingsone, John, 126
Heppingstone v. Mammen, 106, 115, 119, 124, 128–9, 142, 163
Heppingstone, John, 127, 129, 131, 141, 143
Hercules, 116, 118, 129, 142
Hillman, 86, 91, 93, 98–9, 102, 139
Hobbes, Thomas, 166
Hogarth v. Jackson, 53–4, 56, 59, 67
Holmes, Oliver Wendell, 121
Holt, John R., 92, 95, 98, 102, 105
Honolulu, 4, 100, 153
Horwitz, Morton, 59, 104
Howard, Benjamin, 155
Howland, Abraham H., 155
Howland, Isaac C., 116, 129, 131
Howland, Matthew, 155
Hudson, Henry, 37
Hume, David, 165–6, 168
Hunter, 92
Hussey, Christopher, 71
Hutchison v. The Dundee Union Whale Fishing Company, 53

instrumentalism, 103
interference with trade, 55–6
iron holds the whale, 3, 29, 49–50, 56–8, 69, 79–80, 82, 115, 119, 124, 133–4, 168

Jackson, Gordon, 38
Jefferson, Thomas, 149
Jenney, Levi, 93, 96
John, 36
Jonas Island, 84, 122
Jones, John Paul, 73
Justinian, 2, 47, 49, 97–8, 105, 113

Kamchatka, 83–4
Kashevarov Bank, 84
Keeble v. Hickeringill, 55
Kent, James, 59, 62
Krier, James, 171
Kuril Islands, 83

Lamarck, Jean-Baptiste, 151
Lambert, Samuel, 66
lance, 25, 113
lard oil, 153
law-and-economics, 160, 168
law-and-society, 160
legal centralism, 5, 8, 57
legal realism, 103
Leonard, John, 90, 99, 102, 105
Letters of Marque and General Reprisal, 75
Lindholm, Otto, 84–5
Littledale v. Scaith, 36, 42
Living Age, The, 156
Llewellyn, Karl, 102
Locke, John, 57, 163, 166, 171
London, 73, 80
Lowell, John, 104, 106, 110, 112–13, 115, 119–20, 142
Luce, Captain, 77–8
Lushington, Stephen, 100

Mackenzie, George, 43
Malkin, Benjamin Heath, 52, 54
Mammen, John, 125, 127, 141, 143
Manchester, George, 116
Mandeville, Bernard, 166
Mansfield, 1st Earl, William Murray, 45, 60
Maria, 72
Martha, 93
Marvin, William, 101
Massachusetts, 86, 91, 102
mastodons, 149
mateship, 88, 102, 132
McEvoy, Arthur, 13
Melville, Herman, 4–5, 42, 56, 58, 81–2, 105, 147, 163
Menzies, Judge, 67
Middleton, Charles Theodore, 41–2
Milford Haven, 80
Mississippi River, 101
Mitchell, Lawrence, 129
Moby-Dick, 4, 147
Monticello, 139
Montreal, 67
Moody, William, 52, 54
Morgan, Charles W., 149

Nantucket, 4, 69–72, 80, 116
Nantucket Inquirer, 148
natural order, 166, 168
New Bedford, 4, 9, 72, 93, 97, 116, 147, 153, 155
New Bedford Coal Oil Company, 155
New York Evening Post, 156
New York Journal of Commerce, 156
New Zealand, 80
Newfoundland, 71
Nickerson, Asa, 67
Nicoll v. Burstall, 53
norms, 9–10, 13, 132
North East Gulf, Okhotsk Sea, 84, 115, 137
Norton, Hiram, 109
Nye, George, 93

Oil Creek Valley, PA, 154
Okhotsk Sea, 18, 84
Old Middleton, 49
Olmsted, Francis Allyn, 119
Oregon, 125, 127–9
Otis, James, 68

Pacific Commercial Advertiser, 153
Paine, Robert Treat, 68
Parker, Avery F., 92, 95–9, 102, 104–5, 143
Parliament
  Convoy Act of 1798, 75
  Fishery Bill of 1786, 74
Payne, Joseph, 51, 54
Peru, 75
petroleum, 145, 154, 156
  savior of whales, 158
Phoenix, 53
Pierce, Abraham W., 117, 120, 126
Pierson v. Post, 116, 160
Plymouth, MA, 63–4
Presser, Stephen, 103
Priscilla, 42
Prisoner's Dilemma, 134
privity of contract, 67

Quesnay, Francois, 168

R. G. Dun & Co., 155
Rainbow, 116, 118–19, 142
Rattler, 75
res nullius, 41, 97
Resolution, 29, 31–2, 36, 56
Richmond, 123, 127–8
right whales, 17, 70–1, 79, 83, 134, 151
Robertson, George, 123, 128
Rogers, Jonathan, 124, 129, 131
Roman law, 40, 43–4, 48–9, 57, 59–60, 66, 97, 117
Rose, Carol, 161–2
  crystals and mud, 164
Rotch family, 76
Rotch, Francis, 73
Rotch, William, 71
royal fish, 63, 65, 100
Roys, Thomas, 135

Sag Harbor, 4, 18, 108, 112
salvage, 64, 97, 101, 104, 106
Samuel Enderby and Sons Company, 73, 75
Schelling, Thomas, 170
Schooner Reeside, The, 117
Scientific American, 156
Scoresby, William, Jr., 30, 32, 34, 36, 42, 49, 56–7, 82, 90, 131–2, 148, 163
Scoresby, William, Sr., 36
Scottish law, 44, 49
Shantar Bay, 88
Shantar Islands, 83, 85–6, 137, 139
Shasta County, CA, 8
shore whaling, 38–9, 64–5, 67
Silvey, George, 116, 118, 129, 131
Skinner v. Chapman, 53–4, 56, 61
Smith, Adam, 10, 166
Smith, Barzillai, 69
Smith, Gamaliel, 69
Smith, Henry E., 161
Smith, John, 91, 95–6
South Seas Company, 74
Southampton, NY, 63
Southwest Bay, 107
Sowle, J., 125, 131
sperm oil, 17, 72–3, 83, 153
sperm whales, 2, 14–15, 17–18, 20, 25, 28, 41, 69–72, 75, 81, 133–4, 146, 148–9, 151, 158
  behavior, 17, 79
  diet, 15
  population, 16, 83
  range, 16, 83
  stocks, 144
spermaceti, 17
Spitzbergen, 34, 37–8, 79, 114
spontaneous order, 166–7, 172
Sprague, Peleg, 96–7, 100, 102, 105, 110, 112, 119, 139, 141
Stackpole, Edouard, 76
Stair, First Viscount, 41

Starbuck, Alexander, 72
States-General of Holland and West Friesland, 39
Stetson, T. M., 112
Stokes, Whiley, 120
Story, Joseph, 59–60, 112, 117–18
Strait of Belle Isle, 67, 71
Sturges, Thomas, 67
Sugden, Robert, 172
Superior, 18, 83, 135
Supreme Court of Hawaii, 122
Supreme Court of Judicature of New York, 126
Swan, 100
Swan, The (Admiralty case), 100–1
Swift & Perry, 116
Swift v. Gifford, 107, 115, 121, 124, 126, 129, 142

Taber and Company, 93, 96–7, 104
Taber v. Jenny, 86, 92, 105–6, 110–14, 131, 139, 141
Taber, Captain, 91
Taber, Henry, 143
Tasmania, 80
  legislation, 80
The King's Frigate against a Scots Caper, 41
Thomas, Gideon, 64
Thompson, E. P., 13, 47
Thurlow, Lord, 42
tight-knit communities, 161
Tooker, George S., 93, 95, 102, 105
Tortoise, 41
tragedy of the commons, 13, 134, 146, 158, 166
transaction costs, 7–8, 168
Trebatius, 40, 60
Trident, 91
trover, 54–5
turpentine, 152

Udskaya Gulf, 85
Union of 1707, 42

Vanity Fair, 156

waif, 99
Wakefield, Edward Jerningham, 80
Walton, R. Brent, 10
Walvis Bay, 72
Washington, 141
welfare maximization, 8, 10, 168
West Indies, 71
West, Paul, 76
whale bone, 154
whale oil, 21, 72–3, 147, 153–4, 157
  lubricants, 153
whaleboats, 23, 85, 87
whalebone, 20, 22, 28, 83
whalefisheries
  Davis Strait, 67
  Greenland, 4, 29, 32, 36, 38, 40–2, 45, 49, 54, 56–7, 61, 70, 73, 79, 91, 96, 114, 131, 162, 168
  North Pacific, 83, 134–5, 140
  Okhotsk Sea, 83, 85, 87, 94, 99, 106–7, 113–15, 121, 132, 134–5, 143, 162, 168, 171
  Southern, 45–6, 61, 69, 73, 76–8, 80, 114, 132, 134
  Western Arctic, 4, 133, 135, 138, 140
Whalemen's Shipping List, 140, 142, 156
whales
  taxonomy, 14
whaling
  discipline, 109
  hunting techniques, 27, 81
  profits, 22, 134
  techniques, 134, 140
  trying out, 28
whaling techniques, 87
Whitby, 29, 31
White, Herbert, 112
Whitney, George, 90
Wilbur, F. W., 125, 129
Wilcox, P. S., 124–5
Wilkes, Charles, 151
William Fennings, 46, 77
Williams, Thomas, 125–6
Wilson, Richard E., 141
Wrangel Island, 137

Young v. Hitchens, 61

Zone, 92–3, 98, 101, 110, 139